Employer Branding for Competitive Advantage

Information Technology, Management and Operations Research Practices

Series Editors: Vijender Kumar Solanki, Sandhya Makkar, and Shivani Agarwal

This new book series will encompass theoretical and applied books and will be aimed at researchers, doctoral students, and industry practitioners to help in solving real-world problems. The books will help in the various paradigms of management and operations. The books will discuss the concepts and emerging trends in society and businesses. The focus is to collate the recent advances in the field and take the readers on a journey that begins with understanding the buzz words like employee engagement, employer branding, mathematics, operations, technology, and how they can be applied in various aspects. It walks readers through engaging with policy formulation, business management, and sustainable development through technological advances. It will provide a comprehensive discussion on the challenges, limitations, and solutions of everyday problems like how to use operations, management and technology to understand the value-based education system, health and global warming, and real-time business challenges. The book series will bring together some of the top experts in the field throughout the world who will contribute their knowledge regarding different formulations and models. The aim is to provide the concepts of related technologies and novel findings to an audience that includes specialists, researchers, graduate students, designers, experts, and engineers who are occupied with research in technology, operations, and management related issues.

Performance Management
Happiness and Keeping Pace with Technology
Edited by Madhu Arora, Poonam Khurana, and Sonam Choiden

Soft Computing Applications and Techniques in Healthcare
Edited by Ashish Mishra, G. Suseendran, Trung-Nghia Phung

Employer Branding for Competitive Advantage
Models and Implementation Strategies
Edited by Geeta Rana, Shivani Agarwal, and Ravindra Sharma

Analytics in Finance and Risk Management
Edited by Sweta Agarwal, Nidhi Malhotra, and T. P. Ghosh

For more information about this series, visit: www.routledge.com/Information-Technology-Management-and-Operations-Research-Practices/book-series/CRCITMORP

Employer Branding for Competitive Advantage

Models and Implementation Strategies

Edited by

Geeta Rana, Shivani Agarwal, and Ravindra Sharma

CRC Press
Taylor & Francis Group
Boca Raton London New York

CRC Press is an imprint of the
Taylor & Francis Group, an informa business

First edition published 2021
by CRC Press
6000 Broken Sound Parkway NW, Suite 300, Boca Raton, FL 33487-2742

and by CRC Press
2 Park Square, Milton Park, Abingdon, Oxon OX14 4RN

Library of Congress Cataloging-in-Publication Data
Names: Rana, Geeta, editor. | Agarwal, Shivani, editor. | Sharma, Ravindra, editor.
Title: Employer branding for competitive advantage : models and
implementation strategies / edited by Geeta Rana, Shivani Agarwal, and
Ravindra Sharma.
Description: Boca Raton : CRC Press, 2021. |
Series: Information technology, management and operations research practices |
Includes bibliographical references and index.
Identifiers: LCCN 2020045084 (print) | LCCN 2020045085 (ebook) |
ISBN 9780367650964 (hardback) | ISBN 9781003127826 (ebook)
Subjects: LCSH: Employees–Recruiting. |
Branding (Marketing)–Management. | Social media.
Classification: LCC HF5549.5.R44 E67 2021 (print) |
LCC HF5549.5.R44 (ebook) | DDC 658.3/111–dc23
LC record available at https://lccn.loc.gov/2020045084
LC ebook record available at https://lccn.loc.gov/2020045085

ISBN: 978-0-367-65096-4 (hbk)
ISBN: 978-1-003-12782-6 (ebk)

Contents

Preface

With the advent of globalization, organizations today are seen to be increasingly operating in the global village. The struggle for limited resources at the global level seems to be a perennial concern being faced by organizations. This has made the completion between organizations, to survive and excel in the market, more acute than ever before. Successful organizations systematically design, integrate, and proactively implement programs that build and sustain a high-performance workforce. It has become essential for organizations to have a proper system in place that can deal with achieving results for, with, and through people. Talented-motivated employees are a company's best assets, and the techniques in this book help attract, recruit, and retain the very best. A successful employer brand reaches beyond the boardroom to establish confidence, loyalty, and enthusiasm all the way down the ladder. *Employer Brand Management* gives readers a personal grasp of a new approach to people management. It draws on significant advances in practices among leading companies to provide a handbook for employer brand development and implementation. The need of the present book is a result of curiosity to learn about the various innovative practices carried out by the employers to win the battle of being the "Best Employer" in the industry. The book explains how and why organizations should concentrate on their employer brand image to attract and retain the best employee to win the war of talent. Thus, the book gives a well-knitted and balanced coverage of theory, contemporary issues, and practical examples and anecdotes drawn from the Indian business world. This book is an anthology of 12 research papers that provide a rich repertoire of tools and techniques across business functions researched, tested, and validated in various business settings. Human resource agility, corporate social responsibility, talent orientation, corporate brand, digital management of employer brand, role of social media, and branding through workforce techno- innovation in employer branding are the main topics of the book. This book will also enable the readers to learn how to use multi-functional area tools, techniques, innovative frameworks, practices, and approaches for understanding, assessing, and managing the strategic value drivers of business excellence. Overall, this book brings forth a new stream of thoughts by a few fine researchers in the domain of business management. We wish and hope that this book will generate interest among not only fellow academicians but also management practitioners.

Happy Reading!

Editors

Geeta Rana, Shivani Agarwal, and Ravindra Sharma

Acknowledgments

As the proverbial saying, "it takes a village to raise a child" indicates, it involves efforts of several people to bring up a book. This book, as no exception, also grew with the help, intellectual inputs, and efforts of very many people. We take this opportunity to thank all those individuals who have bestowed this book with their work, ideas, time, and valuable inputs.

We gratefully recognize and appreciated the contributions of the authors, insightful reviews and feedback of our panel of reviewers and the patient and concerned efforts of the editorial staff in shaping this book.

We thankfully acknowledge all the support, inspiration, and motivation received from our faculty colleagues. We would remain indebted to their outstanding intellectual efforts. We would also express our gratitude to Dr. Vijay Dhasmana, the Vice Chancellor of Swami Rama Himalayan University, Jolly Grant, Dehradun; and Prof. Alok Saklani Dean of Management, Swami Rama Himalayan University, Jolly Grant, Dehradun, India, for their faith, trust, and responsibility that empowered us and extended full support and for being a constant source of inspiration and pillars of strength throughout the pursuit.

Our Special thanks go to Prof. Renu Rastogi and Prof. Santosh Rangaker of Indian Institute of Technology, Roorkee, India, for their constant encouragement and support in this endeavor. The book in its present shape has been made possible due to keen interest shown by academic colleagues from around the world. We acknowledge their tremendous support and are thankful to them for their valuable comments.

We express our thankfulness to our publisher CRC (Taylor & Francis Publication) and the entire editorial team who lent their wonderful support throughout and ensured timely processing of the manuscript and bringing out this book.

Finally, we would like to give special thanks to our families for unstained support provided while compiling this volume.

Editors

Geeta Rana, Shivani Agarwal, Ravindra Sharma

Acknowledgments

[illegible]

Editors

Geeta Rana is an associate professor in Himalayan School of Management Studies at Swami Rama Himalayan University, Jolly Grant, Dehradun, India. She has earned her PhD from the Indian Institute of Technology (IIT), Roorkee) in human resource (HR) management and organizational behavior. She has also done certification course in HR analytics from the Indian Institute of Management Rohtak (IIMR). She is engaged in teaching, research, and consultancy assignments. She has more than 15 years of experience in teaching and in handling various administrative as well as academic positions. She has to her credit over 40 papers published in refereed journals of Emerald, Sage, Springer, Taylor & Francis, Elsevier, and Inderscience. She also presented several research papers in national and international conferences. Dr. Rana has contributed many chapters in different books published by Cambridge UK, Springer IGI Global, and Palgrave Macmillan. She has authored several books with Taylor & Francis, Nova Publisher, and New Age Publishers. She has conducted and attended various workshops, faculty development programs (FDPs), and management development programs (MDPs) in various institutions. Dr. Rana is a recipient of many reputed awards. Her research interests include knowledge management, managerial effectiveness, justice, values, employer branding, innovation, artificial intelligence, and HR management.

Shivani Agarwal is an assistant professor in KIET School of Management, KIET Group of Institutions, Delhi-NCR, India. She has received her PhD in organizational behavior (Psychology) from the Indian Institute of Technology (IIT), Roorkee. Prior to her current role, she was associated with the Institute of Technology and Science, Ghaziabad, Uttar Pradesh (UP), India; HRIT Group of Institutions, Ghaziabad, UP, India; IIT Roorkee, Uttarakhand, India; and the Centre for Management Development, Modinagar, UP, India. She has attended several short-term courses at IIT Roorkee and IIT Delhi, India, and earned National Programme on Technology Enhanced Learning (NPTEL) certificate for research writing and HR management.

She has authored or coauthored more than 10 research articles that are published in journals, books, and conference proceedings. She teaches graduate and postgraduate level courses in management.

She is the Book Series Editor of Information Technology, Management and Operations Research Practices, CRC Press, Taylor & Francis Group, USA and the guest editor with IGI Global, USA.

Ravindra Sharma is an Assistant Professor in Himalayan School of Management Studies at Swami Rama Himalayan University, Dehradun, India. He has more than 15 years of corporate and academic experience. He holds degrees in Master of Business Administration (MBA) and Master of Computer Applications (MCA). He also qualified University Grants Commission - National Eligibility Test,(UGC-NET). He has organized and conducted a number of workshops, summer internships, and expert lectures for students as well as faculty. He has published several research papers in referred journals of Emerald, Sage, Springer, IGI Global, and Inderscience. He has published books in the area of Internet of Things (IoT) and employer branding with reputed publishers like Taylor & Francis Group, Nova Science Publishers and IRP Publication House. Mr. Sharma has contributed chapters in different books published in Springer IGI Global, Tailor & Frances, and Palgrave Macmillan. He has attended various workshops, FDPs, and MDPs. He has presented several research papers in national and international conferences. He has also attended a paper development workshop in the Indian Institute of Management Rohtak, India. He has been honored as a session chair and keynote speaker in international conferences in various universities. His research interests include IoT, employer branding, entrepreneurship, and talent management.

Contributors

Shivani Agarwal is an assistant professor in KIET School of Management, KIET Group of Institutions, Delhi-NCR, India. She has earned her PhD in organizational behavior (Psychology) from the Indian Institute of Technology (IIT), Roorkee. Prior to her current role, she was associated with the Institute of Technology and Science, Ghaziabad, Uttar Pradesh (UP), India; HRIT Group of Institutions, Ghaziabad, UP, India; IIT Roorkee, Uttarakhand, India; and the Centre for Management Development, Modinagar, UP, India. She has attended several short-term courses at IIT Roorkee and IIT Delhi, India, and earned National Programme on Technology Enhanced Learning (NPTEL) certificate for research writing and HR management.

She has authored or coauthored more than 10 research articles that are published in journals, books, and conference proceedings. She teaches graduate and postgraduate level courses in management. She is the Book Series Editor of Information Technology, Management and Operations Research Practices, CRC Press, Taylor & Francis Group, USA and the guest editor with IGI Global, USA.

Anjuman Antil is a doctoral scholar at the Faculty of Management Studies (FMS), University of Delhi, India. She has about four years of corporate experience and teaching experience in the field of marketing. She has published her papers in journals like *Global Business Review* and *Journal of Creative Communications*. Her research interests include usage of metaphors in communication, branding, marketing, and politics.

Megha Bharti is a doctoral scholar at the Faculty of Management Studies (FMS), University of Delhi, India. She has over a year of teaching experience in the field of economics. She has published many papers in revered journals of management and presented in international conferences. Her research interests include consumer behavior, luxury consumption, conspicuous consumption, and social media branding.

Mostapha El Idrissi holds a doctoral degree in strategic management from the National School of Business and Management of Casablanca, Hassan II University, Morocco. He is a professor of management and entrepreneurship at European Secretariat for Cluster Analysis (ESCA) Business School, Casablanca. His current research interests include collaborations between competitors "co-opetition," small and medium enterprises (SME), open innovation, proximity, and innovation.

Younès El Manzani holds a doctoral degree in management studies from the Jean Moulin Lyon 3 University and the Cadi Ayyad University, Lyon, France as part of a joint PhD. He is an assistant professor at the Institut Supérieur de Management (ISM-IAE), Versailles Saint-Quentin en Yvelines University. Attached to the research laboratory LAREQUOI, his research addresses topics related to strategic management, quality management, innovation management, and entrepreneurship.

Jeevan Jyoti is a senior assistant professor in the Department of Commerce, University of Jammu, Jammu and Kashmir, India. Her areas of research interest are smart human resource management (SHRM), organizational behavior, cross-cultural management, and entrepreneurship. She has publications in eminent journals and edited books, namely, *IIMB Management Review*, *Vision – The Journal of Business Perspective, Journal of Services Research, Annals of Innovation and Entrepreneurship, International Journal of Management Science, Nice – The Journal of Business Perspective, The Indian Journal of Commerce, Metamorphosis,* and *Global Business Review*. She is currently working on SHRM, talent management, outsourcing, and mentoring.

Pavnesh Kumar has done MBA from Faculty of Management Studies, Banaras Hindu university, UP, India with specialization in international business, economics and financial management, and PhD from VBS Purvanchal University, UP, India in rural banking, having a rich academic experience of around 18 years in reputed universities like Institute for Integrated Learning in Management (IILM), ICFAI, and Indira Gandhi National Tribal University (IGNTU). He is currently positioned as professor, head of Department of Management Sciences in Mahatma Gandhi Central University, Bihar. He has published more than 20 research papers in national and international journals in multidisciplinary areas.

Zakaria Lissaneddine holds a PhD in HR management from Cadi Ayyad University Marrakesh, Morocco. He is a member of New Practices of Management Laboratory at the Faculty of Legal, Economic and Social Sciences of Marrakech. His research activities are specifically about digital marketing, social media, and qualitative methods.

Harsh Mishra is an assistant professor in the Department of Journalism and Mass Communication, School of Journalism, Mass Communication and New Media at Central University of Himachal Pradesh. He holds a PhD in journalism and mass communication from the University of Lucknow. He has a decade of teaching and academic research experience. His teaching and research interests include corporate communications, advertising, and journalism studies. He has published with some of the most reputed academic journals. Prior to joining the academics, he has rendered his services as a Corporate Communications Manager to Bank of India.

Zeba Naz is currently working as a faculty in the Department of Business Administration, Aligarh Muslim University, Kishanganj Centre, Bihar. She has received her PhD and Master in Business Administration from the Department of Business Administration, Faculty of Management Studies and Research, Aligarh Muslim University, India. Her research interests include change management and organizational behavior.

Rinku Raghuvanshi is an associate professor in the Department of Management, Bhartiya Skill Development University (BSDU), Jaipur, India. She is also working as a HR manager at BSDU. She has 14 years of rich experience in the academics and administration. She did her PhD in economics from Dr. B R Ambedkar University, Agra and MBA in HR from AMU Aligarh. Raghuvanshi has published more than

15 research papers and articles in reputed journals. Her published work is mainly on Indian agriculture economy, Goods and Service Tax (GST), HR, and informal learning. She has guided five research scholars and is also the Chief Editor of the *EP Journal of Applied Finance and Economy*.

Geeta Rana is an associate professor in Himalayan School of Management Studies at Swami Rama Himalayan University, Jolly Grant, Dehradun, India. She has earned her PhD from the Indian Institute of Technology (IIT), Roorkee in HR management and organizational behavior. She has also done certification course in HR analytics from the Indian Institute of Management Rohtak (IIMR). She is engaged in teaching, research, and consultancy assignments. She has more than 15 years of experience in teaching and in handling various administrative as well as academic positions. She has to her credit over 40 papers published in refereed journals of Emerald, Sage, Springer, Taylor & Francis, Elsevier, and Inderscience. She has also presented several research papers in national and international conferences. Dr. Rana has contributed many chapters in different books published by Cambridge UK, Springer, IGI Global, and Palgrave Macmillan. She has authored several books with Taylor & Francis, Nova Publishers, and New Age Publishers. She has conducted and attended various workshops, FDPs, and MDPs in various institutions. Dr. Rana is a recipient of many reputed awards. Her research interests include knowledge management, managerial effectiveness, justice, values, employer branding, innovation, artificial intelligence, and HR management.

Roomi Rani is an assistant professor in the Department of Commerce, Higher Education Department, J&K, India. Her doctoral research is on "Antecedents and Consequences of Talent Management Practices." She has attended a number of conferences, at both national and international levels. She has published many papers in international and national journals, namely, *Journal of IMS Group*, Emerald Publication, Inderscience Publishers, and *Saaransh*.

Ramzan Sama is an assistant professor in the School of Business Management, SVKM's Narsee Monjee Institute of Management Studies, Indore. He holds a PhD from Gujarat Technological University, Ahmedabad, Gujarat. He has more than a decade-long teaching experience in eminent management institutes of India. His research papers have been published in ABDC and Scopus indexed journals. His corporate experience spans across India's well-known companies like HDFC Bank, Sun Pharma, and Cadila Pharma.

Aditi Sharma is currently working as an assistant professor in the Department of Business Administration (HPKV Business School), Central University of Himachal Pradesh, Dharamshala, India. She has received her PhD from Panjab University, Chandigarh, India. Her research interests include organizational culture, leadership, organizational effectiveness, and healthcare issues.

Antima Sharma is a research scholar in the Department of Management and pursuing her PhD from Bhartiya Skill Development University, Jaipur, India. She did her master's in finance and HR management from Rajasthan Technical University, Jaipur. She is having working experience in the domain of HR management at both industrial and academic levels. Antima is exploring in her research work and has published four quality research papers in reputed journals. She has published paper mainly on development of an individual through informal learning. Her published work reflects about the impact of informal learning on skills development of an individual, employees, and students.

Jeevesh Sharma is a research scholar in the Department of Business Administration, Manipal University, Jaipur, India. She is presently pursing his research work in corporate social responsibility and sustainability.

Ravindra Sharma is an assistant professor in Himalayan School of Management Studies at Swami Rama Himalayan University, Dehradun, India. He has more than 15 years of corporate and academic experience. He holds degrees in MBA and MCA. He also qualified UGC-NET. He has published several research papers in referred journals of Emerald, Sage, Springer, IGI Global, and Inderscience. He has published books in the area of IoT and employer branding with reputed publishers like CRC/ Taylor & Francis, Nova Science Publishers, and IRP Publication House. Mr. Sharma has contributed chapters in different books published in Springer, IGI Global, Taylor & Francis, and Palgrave Macmillan. He has attended various workshops, FDPs, and MDPs, and has presented several research papers at national and international conferences. He has also attended a paper development workshop in the Indian Institute of Management Rohtak, India. He has been honored as a session chair and keynote speaker in international conferences in various universities. His areas of research interest include IoT, employer branding, entrepreneurship, and talent management.

Sajeeb Shrestha is an associate professor at the Faculty of Management, Tribhuvan University Kirtipur, Nepal. He is currently the head of Department of Marketing, a member of the Management Research Department, and a member of the Research Management Cell (RMC) at Shanker Dev Campus, Tribhuvan University. He has been teaching different disciplines in marketing, research, and strategic management since 2008. His research focuses primarily on knowledge management, talent management, employer branding, brand management, service marketing, and customer behavior. He has published many research articles on marketing and management in various national and international journals and presented research papers on national and international conferences in Nepal and India. Dr. Shrestha has authored some books in marketing like marketing, advertising, distribution management, international marketing, and marketing research.

Shweta Taluka is a research scholar in the Department of Business Administration at Manipal University, Jaipur, India. She is presently pursing her PhD research work is in corporate governance and banking.

Suhasini Verma is presently working as an associate professor in the Department of Business Administration, Faculty of Management and Commerce, Manipal University, Jaipur, India. Her research interests include economics, finance, corporate governance, and financial inclusion.

Chandan Veer has done Master of Business Administration (MBA) in marketing from University of Pune, having a rich corporate experience of food and beverage industry with multinational corporations like United Spirits Limited (UB Group). He is a faculty in the Department of MBA in B.D. College, Patna and is pursuing his PhD from Mahatma Gandhi Central University, Bihar, India.

Farah Zahidi is a faculty in the Department of Business Administration, Aligarh Muslim University, Kishanganj Centre, Bihar. She is also pursuing her PhD in HR management. She has a Master's in Business Administration and is also a law graduate from Aligarh Muslim University, India. Her areas of research interest include organizational behavior and related topics.

1 Modeling Drivers of Employer Branding

Agile Role of HR

Zeba Naz and Farah Zahidi

INTRODUCTION

Even though the concept of branding in human resource management (HRM) is relatively new, the last decade has seen tremendous increase in the importance and application of the same by organizations. The volatile nature of the business environment makes it necessary for the organizations to make efforts toward attracting talented new recruits as well as keeping their employees engaged and internalizing the organization's values. Employer branding has established itself as one of the most important factors of competitive advantage (Conference Board, 2001). Employer branding helps in the integration of the various related yet different constructs with managerial as well as scholarly importance under one umbrella. The contribution of such an umbrella concept toward the quest for the search of establishing a framework for strategic human resource management can be unparalleled (Backhaus and Tikoo, 2004). Organizations build their employer branding quite systematically, which results in financial benefits, an increase in attractiveness to prospective employees, and the creation of a suitable and effective organizational culture (Figurska and Matuska, 2013).

The intense competition in the employment market toward securing as well as retaining a talented work pool has established employer branding as one of the strategically imperative functions (Branham, 2001). Employer branding is basically the association of positive perceptions with any organization by its prospective employees as well as its existing talents. It enhances the organization's strength by creating a value proposition. It helps in aligning different psychological, financial, and functional benefits and positioning the organization as the "employer of choice." However, such associations and perceptions must be clear, consistent, and credible (Wilden, Guderganand Lings, 2010). Employer brand is an asset to the organization which needs to be attended and cultivated regularly (Sharma, Singh and Rana,2019).

MEANING AND DEFINITION

In the traditional sense, the term "branding" has always been associated with the marketing of products. However, with the passage of time, branding has also been associated with the differentiation of firms, places, and people (Peters, 1999). Employer branding provides identity to the unique characteristics of the organizations which helps them in attracting new employees and retention of employees. Employer branding has also been defined by Ambler and Barrow (1996) in the context of benefits. They identified financial, psychological, and functional benefits as the crucial elements of employer branding.

Employer branding is the amalgamation of both HRM and marketing (Wojtaszczyk, 2012). But, various surveys have established a link between strategy and employer branding. Employer branding helps in aligning the trifecta of strategic HRM, a marketing strategy that is competitive, and effective leadership (Minchington, 2006.

Rosethorn (2009) identified employer branding as a mutual relationship between the employer and employee. But the deal between the employer and the employee should be peculiar, captivating, and very much relevant to the prospects. Most importantly, employer branding must be carried out throughout the whole association between the employer and employee. Branding helps in providing people with reasons to choose to join the organization and thereafter stay there. Employer branding helps in establishing an image for the organization. This image creates a positive perception for the organization in the minds of the employees as well as in the external markets (SHRM, 2008). Knox and Freeman (2006) concluded that employer branding is a combination of three things –internal branding (present employees), external branding (stakeholders), and recruiters' (potential employees) perceptions of the firm. Organizations use employer branding as a means to secure access to potential employees (Branham, 2001).

But employer brand is not something that any organization has to create from scratch. Employer brand somehow coexists with the organizational culture. Influence of the strategies used by the organizations also cannot be ignored (RanaetAl., 2016). Hence, the organizations must build this image consciously. It has to create an attractive, reputable, and quality conscious perception of an organization that gives importance to the all-round well-being of its employees along with corporate social responsibility.

In the present scenario, due to the ramifications of the coronavirus disease 2019 (Covid19) pandemic, many organizations have been forced to downsize and lay off their employees. This has taken a hit on the brand status of the concerned organizations, which might have been bruised in the aftermath. Companies need to reverse this course and intensify their efforts toward it. The organizations need to be nimble, proactive, and agile to combat the situation. In order to create a strong and distinctive brand, organizations must identify those qualities that make them distinct and will help in establishing an emotional connection between the employers and employees (Sartain and Schumann, 2006).

LITERATURE REVIEW

Even though the concept of employer branding is still not very old, it has managed to gain and hold the attention of organizations. The concept was introduced by Ambler and Barrow (1996). They tested the application of different brand management techniques in the context of HRM. Functional, economic, and psychological benefits were found to be associated with employer branding. The employer offers them and the employees identify them. Employer branding can be used as a strategy by organizations to differentiate themselves as employers, with the ultimate objective of attracting potential talents as well as existing workforce (Gehrels and de Looij, 2011).

Employer branding is a three-step process: first, organizations create a value proposition, then they market this very proposition to the external customers, and, finally, they make internal marketing of the employer brand, that is, create a workforce committed and loyal to the organization (Backhaus and Tikoo, 2004; Sullivan, 2002; Frook, 2001). Brands must be consistent with the corporate brand and the products being offered by the organization. The intense competition of the business world makes it imperative for employers to make themselves distinct in the eyes of the prospective employees. By virtue of such distinct qualities only an organization would be able to attract and retain a talented work pool (Mosley, 2007). This tenacious need for quality employees has played an important role in establishing employer branding as an essential strategy for organizations (Rana, and Sharma, 2019).

When an organization is able to successfully create an employee brand, it can reap multiple benefits. The employee turnover gets reduced; the satisfaction level of employees increases, customers become more loyal and satisfied, and the reputation among the stakeholders also gets enhanced (Ewing, Pitt, De Bussy and Berthon, 2002). In order to make a clear statement of the standing of the organization, an employer must choose an effective communication channel and all the branding initiatives must be continually measured and evaluated for effectiveness. However, the bottom line is aligning the expectations and ground reality for the employees in the organization. What is being promised must be delivered too (Paul and Kanthimathi, 2016).

Various researches have been conducted to identify the elements critical to employer branding. Efforts were made to understand the areas where focused approach should be made by the organizations to create a valuable employer brand. Sullivan (2004) identified a set of eight elements that can help in establishing a successful employer brand. A sharing culture with continuous improvement, striking appropriate balance between productivity and management style, gaining public recognition, a pool of proactive employees, popularity, setting and achieving benchmarks, increased candidate awareness, and branding assessment metrics were earmarked as the critical elements of a successful employer brand.

Griffin and Clarke (2008) were of the view that it is important for organizations to have an in-depth understanding of their current and aspirational standing. They must have a well-developed plan which they are able to execute successfully. Employee engagement should be successfully implemented along with a regular review of all the plans that have been achieved. Organizations need to identify the needs of the target group and create a unique employer value proposition (EVP) for them (Botha, Bussin and Lukas, 2011). They should follow a people strategy and whatever brand identity is being pitched to the audience must be consistent in nature. A proper brand communication channel must be established and reliable employer branding metrics must bhe used.

Shitika, Tanwar, and Shrimali (2013) zeroed in on 11 different elements that contribute towards creating a successful employer brand. They include company image, attrition, internal recruitment, public relation, employee satisfaction, employee engagement, brand portfolio, work environment, communication channel, strategic policy, and employment offerings. Subramoniam (2019) came out with a different set of drivers which help in establishing a reputed employer brand. He concluded that organizations can create a reliable employer brand if they are able to engage their present employees, taking efforts toward making their employees committed and loyal and finally creating an attractive image for the organization in the prospective employees.

On the basis of literature review some of the critical factors that have found to be major contributors to employer branding are summarized–in Table 1.1.

RESEARCH METHODOLOGY

Inventories of crucial employer branding factors were identified from the literature, which were further validated by taking inputs from HR experts. The factors depicted employee's preference for selecting employers. It was measured by using multiple criteria and then the opinions of experts helped in identifying the interrelationships. In this study, perceptual measure was used. Then a model depicting the relationships and hierarchy of factors was drawn using the interpretive structural modeling (ISM).

ISM is a well-established methodology for identifying relationships among variables. It helps in the conceptualization and theory building, which can be tested later. The model produced portrays a structure that can depict the complexity of a relationship. This method depends on expert's judgments on whether and how the variables are related. The variables are identified from the literature and any suitable group process like brainstorming can be done to get the expert's opinion (Sushil, 2012).

The various steps involved in ISM technique are as follows:

Step 1: Different outcomes (or variables), which are related to defined problems and are identified and enlisted by a survey or group problem-solving technique. After this, a contextual relationship is established among factors with respect to whom the pairs of other factors would be examined.

Step 2: A structural self-interaction matrix (SSIM) is developed for factors. This matrix indicates the pair-wise relationship among employer branding factors. This matrix is checked for transitivity. The contextual relationships among the factors were identified by the experts as shown in Table 1. For example, the experts believed that pleasant work atmosphere will lead to work–life balance (Appendix 1.1).

TABLE 1.1
Critical Factors Impacting Employer Branding

Critical Factors	Meaning	References
Financial Health	A financially healthy organization is able to promise and deliver stability to the employees.	Randstad Global Report (2018)
Strong Management	If the management is able to provide the employees with good leadership support and take decisions that are beneficial for the employee, it increases the employer branding of the organization.	Pahor and Franca (2012), Kudret (2014), and Jain and Pal (2012)
Career Prospects	Organizations offering chances of development throughout the span of career in the organization possess favorable perceptions.	Voronchuk and Stariņeca (2014), Subramoniam (2019), and Wilden, Gudergan and Lings (2010)
Good Training	Good training schedule exposes the employees to the culture of the organizations and positively influences the employees' journey and prepares them for their employment cycle. It further improves the chances of the organization to become the "employer of choice."	Munsamy and Venter (2009), Marchington and Wilkinson (2012), and Taylor (2002)
Job Content	The job characteristics, opportunities, and learning environment provided to the employees add in the attractiveness of the organization to the employees, future or present.	Gaddam (2008) and Bondarouk et al. (2012)
Corporate Social Responsibility	It generates many forms of competitive advantage including attractiveness – increases a business's competitive position and performance. Organizations with good brand image have been found to be following and imbibing CSR practices.	Jain(2013), Mahmoud et al. (2017), and Stuss (2018)
Pleasant Work Atmosphere	The work environment must be free from biases, discrimination, and rigid classifications. No glass ceilings should be defined.	Backhaus and Tikoo (2004), Ind (1997), and Melewar et al. (2012)
Long-Term Job Security	The sense of security related to a job is positively linked to the perceptions of employee branding.	Subramoniam (2019) and Gaddam (2008)

(*Continued*)

TABLE 1.1
Cont.

Critical Factors	Meaning	References
Work–Life Balance	It is reflected in work flexibility policies followed by the organizations. This not only helps in creating a loyal workforce but also makes them more loyal to the organization.	Menor (2010) and Memon and Kolachi (2012)
Salary and Employee Benefits	It is one of the major contributors of a successful employer brand; good salary and benefits increases the attractiveness of the organization in the eyes of the prospective and present employees.	Erlenkaemper, Hinzdorf, Priemuth, and von Thaden (2006), Jain and Pal (2012), and Voronchukand Stariņeca (2014)

Source: Authors' Own Work.

APPENDIX 1.1
Structural Self-Interaction Matrix (SSIM)

S.No.	Employer Branding Factors	10	9	8	7	6	5	4	3	2
1	Salary and Employee Benefits	o	o	x	x	x	x	x	a	a
2	Long-Term Job Security	o	o	o	a	a	x	x	x	
3	Pleasant Work Atmosphere	o	a	a	a	x	x	v	-	
4	Work–Life Balance	a	a	a	a	a	a	-		
5	Financial Health	o	o	o	x	v	-			
6	Career Prospects	o	a	x	v	-				
7	Job Content	o	a	a-	-					
8	Good Training	o	x	-						
9	Strong Management	v	-							
10	Corporate Social Responsibility	-								

Step 3: A reachability matrix (RM) is developed from the SSIM, in which the information of each entry in SSIM is transformed into 0's and 1's (Appendix 1.2).
Step 4: The RM is partitioned into different levels (Appendix 1.3).
Step 5: Based upon the above, a directed graph (diagraph) is drawn and transitivity links are removed (Appendix 1.4).
Step 6: Digraph is converted into an ISM model by replacing nodes of the elements with statements (Figure 1.2).

APPENDIX 1.2
Initial Reachability Matrixes (IRM)

S.No.	Employer Branding Factors	1	2	3	4	5	6	7	8	9	10	Driving Power	Rank Order as per Driving Power
1	**Salary and Employee Benefits**	**1**	**0**	**0**	**1**	**1**	**1**	**1**	**1**	**0**	**0**	6	2
2	**Long-Term Job Security**	**1**	**1**	**1**	**1**	**1**	**0**	**0**	**0**	**0**	**0**	5	3
3	**Pleasant Work Atmosphere**	**1**	**1**	**1**	**1**	**1**	**1**	**0**	**0**	**0**	**0**	6	2
4	**Work–Life Balance**	**1**	**1**	**0**	**1**	**0**	**0**	**0**	**0**	**0**	**0**	3	4
5	**Financial Health**	**1**	**1**	**1**	**1**	**1**	**1**	**1**	**0**	**0**	**0**	7	1
6	**Career Prospects**	**1**	**1**	**1**	**1**	**0**	**1**	**1**	**1**	**0**	**0**	7	1
7	**Job Content**	**1**	**1**	**1**	**1**	**1**	**0**	**1**	**0**	**0**	**0**	6	2
8	**Good Training**	1	**0**	**1**	**1**	**0**	**1**	**1**	**1**	**1**	**0**	7	1
9	**Strong Management**	**0**	**0**	**1**	**1**	**0**	**1**	**1**	**1**	**1**	**1**	7	1
10	**Corporate Social Responsibility**	**0**	**0**	**0**	**1**	**0**	**0**	**0**	**0**	**0**	**1**	2	5
Dependence		8	6	7	10	5	6	6	4	2	2		
Rank order as per dependence		2	4	3	1	5	4	4	6	7	7		

APPENDIX 1.3
Partitioning the Reachability Matrix into Different Levels

Employer Branding Factors	Reachability Set	Antecedents	Intersection	Level
1	1,4,5,6,7,8	1,2,3,4,5,6,7,8	1,4,5,6,7,8	1st level
2	1,2,3,4,5	2,3,4,5,6,7	2,3,4,5	2rd level
3	1,2,3,4,5,6	2,3,5,67,8,9	2,3,5,6	2nd level
4	1,2,4	1,2,3,4,5,67,8,9,10	1,2,4	1st level
5	1,2,3,4,5,6,7	1,2,3,4,7	1,2,3,4,7	7th level
6	1,2,3,4,6,7,8	1,3,5,6,8,9	1,3,6,8	5th level
7	1,2,3,4,5,7	1,5,6,7,8,9	1,5,7	3rd level
8	1,3,4,6,7,8,9	1,6,8,9	1,6,8,9	4th level
9	3,4,6,7,8,9,10	8,9	8,9	6th level
10	4,10	9,10	10	3rd Level

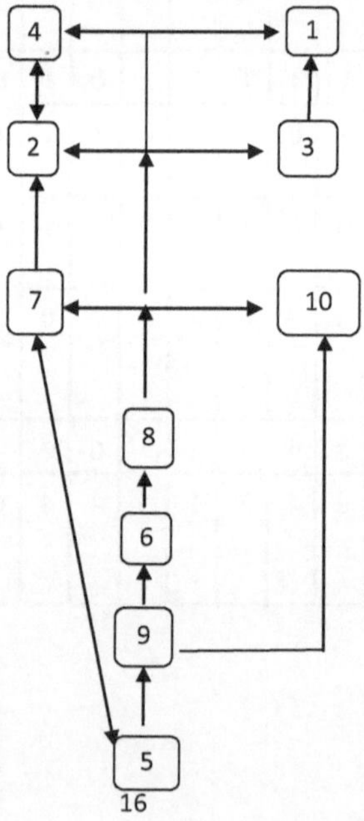

APPENDIX 1.4 Diagraph.

Step 7: Finally, the ISM model is checked for conceptual inconsistency and necessary modifications are incorporated. For analyzing the driving power and dependence of change outcomes, it has been classified into four categories using MICMAC (Matrice d'Impacts croises-multipication appliqué an classment which means cross-impact matrix multiplication applied to classification) analysis on the basis of driving power and dependence.

Divided into four clusters (Figure 1.3), the first cluster includes autonomous factors having weak driving power and weak dependence. The factors in this quadrant are relatively disconnected from the study. Corporate social responsibility that seems to be disconnected from other factors is shown in this quadrant. The second cluster consists of the dependent variables with weak driving power but strong dependence. In our study, factors like salary and employee benefits and work–life balance fall in the second quarter. In the third cluster, the variables have linkage effect with strong driving power and also strong dependence. In the present study, factors like long-term job security, pleasant work atmosphere, job content, and good training fall in this cluster. Any action on these variables will affect others and there will be a feedback effect on them that makes them unstable in the system. Finally, the fourth cluster is called as independent variable, with low dependence and high driving power. In our study, factors like financial health, strong management, and career prospects fall in this category. The factors falling in the third and fourth clusters are called as key variables (Appendix 1.5).

Discussion

Based on the inputs from the experts the ten identified factors of employer branding were partitioned into seven levels. Financial health has emerged as the basic and

APPENDIX 1.5

	10			5							
	9										
	8			9				2	3		
	7		**IV**		6		7		**III**		
	6					8					
Driving Power	5										
	4								**II**		
	3		**I**					1		4	
	2										
	1	10									
		1	2	3	4	5	6	7	8	9	10

Dependence

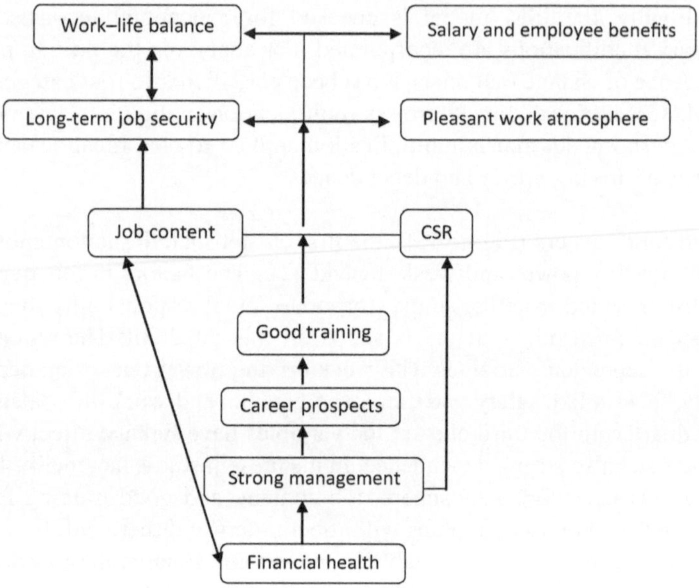

FIGURE 1.1 ISM model (Figure depicting hierarchical and interrelationship among ten factors of employer branding).

(Source: Authors' own work.)

strongest driver of increasing employer branding. Having a strong management and good career prospects are also essential factors as both emerged as the drivers in the hierarchical model developed. Good training, job content, long-term job security, and pleasant work atmosphere fall in the middle of the model, which depicts that they have less driving and dependence power. These factors lead to achieving work–life balance and salary and employee benefits, both having high dependence on other factors. Corporate social responsibility seems to be having less impact on other factors as it emerged as a weak driver and weak dependent variable. The hierarchical model in Figure 1.1 depicts how each factor is interrelated to and impacts each other.

CONCLUSION

Employer branding has emerged to be an important concept for HR managers. The factors impacting it need to be analyzed through a deeper lens. However, the present study suggested a model depicting their inter relationships and dependency, which can prove to be helpful for framing strategies to retain and attract employees. The dimensions mentioned in this study may not happen to the same extent and thus a relative emphasis on any of the dimensions might have a significant negative impact on the employer branding process. For example, an emphasis on good training may result in satisfied and efficient employees but weak management may not be able to utilize that potential, while an emphasis on career prospect can lead to the achievement of individual goals but with a poor job content employees will lack motivation. However, every factor should be given importance but they may probably occur at different speeds. Adopting multiple perceptual measures would facilitate a more holistic view of employer branding process that takes account of desired stakeholders.

As mentioned earlier, the model developed may provide a suitable roadmap for managers to make their employer brand strong by cautiously considering each factor. However, the model could not be generalized to other settings as the inputs are taken from HR managers of one organization only. A detailed study can be conducted in future for different sectors for enhancing the credibility of the model. The findings of this study were also communicated to the HR managers for any additional input and their views on the hierarchical relationships after conducting ISM. Other organizations can also use this model for enhancing their brand image by forming strategies that could be tailor-made according to their employees.

REFERENCES

Agarwal, S., Jindal, A., Garg, P. and Rastogi, R. (2017). The influence of quality of work life on trust: empirical insights from a SEM application. *International Journal of Indian Culture and Business Management*, 15(4), 506–525.

Ambler, T. and Barrow, S. (1996). The employer brand. *Journal of Brand Management*, 4, 185–206.

Backhaus, K. and Tikoo, S. (2004). Conceptualizing and researching employer branding. *Career Development International*, 9(5), 501–517.

Bondarouk, T., Ruël, H. and Weekhout, W. (2012). Employer branding and its effect on organizational attractiveness via the world wide web: results of quantitative and qualitative studies combined. *The 4th International e-HRM Conference "Innovation, Creativity and eHRM,"* March 28–29, Nottingham, UK.

Botha, A., Bussin, M. and Swardt, L. (2011). An employer brand predictive model for talent attraction and retention. *SA Journal of Human Resource Management*, 9(1), 1–12.

Branham, L. (2001). *Keeping the People Who Keep You in Business: 24 Ways to Hang on to Your Most Valuable Talent.* New York, NY: American Management Association.

Erlenkaemper S., Hinzdorf T., Priemuth K. and von Thaden C. (2006). Employer branding through preference matching. In: Domsch M.E. and Hristozova E. (eds.), *Human Resource Management in Consulting Firms.* Berlin: Springer.

Ewing, M.J., Pitt, L.F., De Bussy, N.M. and Berthon, P. (2002). Employment branding in the knowledge economy. *International Journal of Advertising*, 21(1), 3–22.

Figurska, I. and Matuska, E. (2013). Employer branding as a human resources management strategy. *Human Resources Management & Ergonomics,* VII, 35–51.

Frook, J.E. (2001). Burnish your brand from the inside. *B to B*, 86, 1–2.

Gaddam, S. (2008). Modelling employer branding communication: the softer aspect of HR marketing management. *ICFAI Journal of Soft Skills*, 2(1), 45–55.

Griffin, L. and Clarke, T. (2008). Employer branding. Your customers know your brand & values… do your employees? *Bridge Partners Insights*, August 2008.

Ind, N. (1997). *The corporate brand.* Basingstoke: Macmillan Business.

Jain, S. (2013). Employer branding and its impact on CSR, motivation, and retention of employees using structural equation modelling. *Delhi Business Review*, 14, 83–98.

Jain, V. and Pal, R. (2012). Importance of employer branding in business upgradation. *International Journal of Research in IT & Management*, 2(11), 68–75.

Knox, S., and Freeman, C., (2006), "Measuring and managing employer brand image in the service industry", Journal of Marketing Management, 22, 695-716.

Kudret, S. (2014). Branding the employment experience. *The ESCP Europe 9th International Marketing Trends Conference*, January 24–25, Venice, Italy.

Mahmoud, M.A., Blankson, C. and Hinson, R.E. (2017). Market orientation and corporate social responsibility: towards an integrated conceptual framework. *International Journal of Corporate Social Responsibility*, 2, 9.

Marchington, M. and Wilkinson, A. (2012). *Human Resource Management at Work*, 5th edn. London: Chartered Institute of Personnel and Development.

Melewar, T.C., Gotsi, M. and Andriopoulos, C. (2012). Shaping the research agenda for corporate branding: avenues for future research. *European Journal of Marketing*, 46(5), 600–608.

Memon, M.A. and Kolachi, N.A. (2012). Towards employees branding: a nexus of HR and marketing. *Interdisciplinary Journal of Contemporary Research in Business*, 4(2), 4–61.

Menor, J.H. (2010). 10 Strategic tips for employee retention. *The Recruiters Lounge*. Available at: www.therecruiterslounge.com/2010/08/17/10-strategic-tips-foremployee-retention/.

Minchington, B. (2006). Measuring employer brand effectiveness. Available at: www.pageuppeople.co.uk/Newsletter_Dec2006.htm.

Mosley, R.W., (2007), "Customer experience, organisational and the employer brand", Brand Management, 15(2), 123–134.

Munsamy, M. and Venter, A.B. (2009). Retention factors of management staff in the maintenance phase of their careers in local government. *South African Journal of Human Resource Management*, 7(1), 187–195.

Pahor, M. and Franca, V. (2012). The strength of the employer brand: influences and implications for recruiting. *Journal of Marketing and Management*, 3(1), 78–122.

Paul, M.T.V and Kanthimathi, S. (2016). Anconceptual study on employer branding in Indian organizations. *International Journal of Applied Research*, 3, 861–865.

Peters, T. (1999). *The Brand You 50: Fifty Ways to Transform Yourself from an Employee into a Brand That Shouts Distinction*. New York, NY: Knopf Publishers.

Rana, G., Rastogi, R. and Garg, P. (2016). Work values and its impact on managerial effectiveness: a relationship in Indian context. *Vision*, 22, 300–311.

Rana, G. and Sharma, R. (2019). Assessing impact of employer branding on job engagement: a study of banking sector. *Emerging Economy Studies*, 5(1), 7–21. https://doi.org/10.1177/2394901519825543

Randstad. (2018). *Employer brand search global report*. Available at: https://workforceinsights.randstad.com/global-employer-brand-research-2018#anchorcountryreports

Rosethorn, H. (2009). *The Employer Brand. Keeping Faith with the Deal.* Farnham: Gower Publishing Limited.

Sartain, L. and Schumann, M. (2006). *Brand from the Inside*. San Francisco, CA: John Wiley & Sons.

Shitika, Tanwar, S. and Shrimali, V. (2013). Modelling effectiveness of employer branding – an interpretive structural modelling technique. *Pacific Business Review International*, 5(11), 1–7.

SHRM. (2008). The employer brand: a strategic tool to attract, recruit and retain talent. *Society for Human Resources Management*. April/June 2008.

Sjoerd, A.G. and Joachim de, L. (2011). Employer branding: a new approach for the hospitality industry. *Research in Hospitality Management*, 1(1), 43–52.

Stuss, M. (2018). Corporate social responsibility as an employer branding tool: the study results of selected companies listed on GPW. *International Journal of Contemporary Management*, 17(1), 249–267.

Subramoniam, R. (2019). An ISM based employer branding model – in case of a KPO. *Journal of Xi'an University of Architecture & Technology*, XI(XII), 512–1519.

Sullivan, J. (2002). Crafting a lofty employment brand: a costly proposition. *ER Daily*, November 25.

Sullivan, J. (2004). Eight elements of a successful employment brand. *ER Daily,* February 23.

Sushil (2012). Interpreting the interpretive structural model. *Global Journal of Flexible Systems Management*, 13(2), 87–106.

Tanwar, K. and Prasad, A. (2016). Exploring the relationship between employer branding and employee retention. *Global Business Review*, 17(3_suppl), 186–206.

Taylor, S. (2002). *The Employee Retention Handbook.* London: Chartered Institute of Personnel and Development.

(The) Conference Board. (2001). *Engaging Employees through Your Brand*. New York, NY: The Conference Board.

Voronchuk, I. and Starineca, O. (2014). Human resource recruitment and selection approaches in public sector: case of Latvia. International Scientific Conference *"New Challenges of Economic and Business Development – 2014"* Proceedings, 2014, 417–430.

Wilden, R., Gudergan, S. and Lings, I. (2010). Employer branding: strategic implications for staff recruitment. *Journal of Marketing Management*, 26(1), 56–73.

Wilden, R., Gudergan, S. and Lings, I. (2010). Employer branding: strategic implications for staff recruitment. *Journal of Marketing Management*, 26(1–2), 56–73.

Wojtaszyk, K. (2012). *Employer Branding CzyliZarządzanieMarkąPracodawcy*. Łódź: Wyda wnictwoUniwersytetuŁódzkiego.

2 Assessing the Validity of Employer Branding and Predicting Its Talent-Oriented Outcomes

An Employee's Perspective

Jeevan Jyoti and Roomi Rani

INTRODUCTION: EMPLOYER BRANDING AS A NEW DIMENSION

The concept of employer branding (EB) has been emerged in the 1990s. Since then, it has become the most popular human resource (HR) term, which is widely adopted by managements at global level (Little, 2010). It helps to gain a strong position in the competitive labor market by creating a clear and reputed image of the organization, which helps to attract high-potential employees for vacant positions and to motivate, engage, and retain its current employees (Priyadarshi, 2011; Mandhanya & Shah, 2010; Srivastava & Bhatnagar, 2010). In this context, EB is a new dimension in the field of modern human resource management (HRM), which is aligned with branding principles for attraction, identification, retention, and engagement of talented employees to run organizations smoothly, effectively, and efficiently (Lockwood, 2010; Priyadarshi, 2011; Srivastava & Bhatnagar, 2010; Mihalcea, 2017; Botha et al., 2011). It has also been labeled as 'talent branding', which acts as a keystone for an efficient long-term retention and recruitment strategy (Wilden et al., 2010). Despite gaining extensive popularity in modern HR literature, empirical research is still relatively limited in this field (Davies, 2008). Furthermore, most of the studies have focused only on one or two practices of talent management, such as talent attraction and talent retention. Research gap also revealed that there is a need to validate EB scale, especially in the banking sector.

OBJECTIVES OF THE STUDY

THE OBJECTIVES OF THE STUDY WERE TO:

1. Develop and validate employer branding scale
2. Find out the impact of employer branding on talent management and its practices.

EMPLOYER BRANDING IN INDIAN CONTEXT

During the 1990s, the liberalization period and by the subsequent economic reforms, Indian companies strategically utilized the EB as a tool for attracting and retaining talented employees, which helped to enhance organizational reputation domestically as well as internationally (Dawn & Biswas, 2011; Sharma et al., 2018). Indian EB is comprehensive and mirrors employee value proposition with core corporate values, which can distinguish Indian companies from those in other countries for competitive advantage (Das & Rao, 2012; Dawn & Biswas, 2011). Effective EB practices affect positively job engagement in the organizations (Rana & Sharma, 2019).

Apart from external branding, Indian organizations have realized that they should also adopt internal branding strategies to leverage upon the employees before taking the brand to the market. It is of no use to spend millions on marketing campaigns or infrastructure for a potential customer if the internal employees are not delighted. Some companies like Wal-Mart, Infosys, TCS, Tata Steel, and CEAT used EB to attract and recruit the best talent, through indication given by their employees in the recruitment advertisements (Dawn & Biswas, 2011). A huge amount is spent by them to create EB in India. Research by Kapoor (2010) revealed that 95% Indian HR professionals are aware of the concept of EB and they play an active role towards achieving it. Srivastava and Bhatnagar (2010) suggested some factors, such as caring organization, enabling organization, career growth, credible, fair, flexible, and ethical products and services brand image, positive employer image, and global exposure, which have significant impact on EB in India as compared to the financial performance, industry/sector image, and word of mouth.

CONCEPTUAL FRAMEWORK AND HYPOTHESES DEVELOPMENT

EB process focuses on attraction, acquisition, identification, planning, development, and retention of talent pools (Jyoti & Rani, 2014; Mahesh & Suresh, 2019; Lockwood, 2010). An effective EB strategy helps to reduce unnecessary recruitment cost due to employee turnover and strengthen employer–employee relationship as compared to organization with weaker/absent EB (Ritson, 2002). EB is also helpful in identifying talented/high-potential employees (Priyadarshi, 2011; Piansoongnern et al., 2011; Mandhanya & Shah, 2010). EB is used as a lens for attracting and identifying the employees. It appeals new and existing employees and ensures that all talented employees are identified and engaged within the organization. On the same line, Moroko and Uncles (2008) stated that EB plays a major role at every phase of employment lifecycle, that is, from initial attraction to identification, employment, and retention. Furthermore, EB spends large amount of money in maintaining career websites, career fairs, internships, and organizing competition on employer-of-choice award to develop their talented employees (Wilden et al., 2010). Edwards (2010) argued that EB is not only used to develop a distinctive reputation in the business market (external), but it is also used by employers (internal) to drive positive employee engagement and retention (Wilden et al., 2010; Ibrahim et al., 2018; Rana et al., 2019). Thus, it is hypothesized that:

H1: EMPLOYER BRANDING SIGNIFICANTLY AFFECTS TALENT MANAGEMENT

H1(a): Employer branding significantly affects talent attraction.
H1(b): Employer branding significantly affects talent identification.
H1(c): Employer branding significantly affects succession planning.
H1(d): Employer branding significantly affects talent development.
H1(e): Employer branding significantly affects talent engagement.
H1(f): Employer branding significantly affects talent retention.

RESEARCH DESIGN AND METHODOLOGY

If an organization has a reputation as a good employer, it will be easy for it to pick up talented employees not only from outside but also from internal customers, that is, employees. Keeping this in mind, our study tries to develop theoretical relationships between employer branding, talent management, and its practices in the banking sector, which was tested using appropriate techniques or methods.

Scale Generation (Measures): The concepts in the review of related literature were used for designing the questionnaire for all constructs. After the questionnaire was designed by the researcher, it was validated by the expert of HRM (guide) and one branch manager of each bank. The final questionnaire was consisted of two sections. Section I was a general information sheet about the respondents, such as name of the bank, designation, qualification, age, work experience in present bank, corporate experience, total experience, monthly salary, and last promotion (in year). Section II was based on seven dimensions along with their items. A five-point Likert scale was used to achieve uniformity, ranging from 1 (strongly disagree) to 5 (strongly agree).

Employer Branding: It is an 11-item scale generated from Vaijayanthi et al. (2011) to measure the bank's reputation as a good employer in terms of fulfilling the basic needs of the employees.

Talent Attraction (TA): This included self-generated nine items that assessed the attractiveness of the bank for both current and upcoming employees using the methods described in Highhouse et al. (2003).

Talent Identification (TI): This dimension consisted of tenitems, which were self-generated after reviewing the following relevant literature: Yarnall (2011), Piansoongnern et al. (2011), Hartmann et al. (2010), Horvathova and Durdova (2010), and Edwards and Bartlett (1983).

Succession Planning (SP): This has been assessed with eightitems, which were generated from Piansoongnern et al. (2011), Hartmann et al. (2010), and Bano et al. (2010).

Talent Development (TD): This has been assessed with12 self-generated items from the following research papers: Yarnall (2011), Bano et al. (2010), and Collings and Mellahi (2009).

Talent Engagement (TE): This scale consisted of tenitems, which was adopted from Lockwood (2007).

Talent Retention (TR): This construct consisted of tenitems after reviewing the related literature on retention of employees, such as Piansoongnern et al. (2011), Hausknecht et al. (2009) and Bhatnagar (2007).

Sample

For selecting the sector, a pilot survey on 20 bank employees (conveniently selected) was done to know the scope of EB and talent management practices in the banking sector. Due to the positive response, the data were collected from four public banks, that is, State Bank of India (SBI), Punjab National Bank (PNB), Bank of India (BOI), and Industrial Development Bank of India (IDBI) in Jammu and Amritsar cities. A total of 400 employees were contacted for primary data collection by convenient sampling method, of which 363 employees responded back properly. The average age of the respondents was about 42 years and the average monthly income was about Rs. 35,303. Majority of employees (53%) had1 to 10 years' work experience in the banking sector. About 67% were assistant managers (scale1) and 33% were workmen staff (clerks). Furthermore, 21.7% of employees were graduates, 28.7% were graduates with Certified Associate of Indian Institute of Bankers (CAIIB) certification, 21.7% were postgraduates, 9.2% had a postgraduation qualification with CAIIB, 16.9% had professional degrees, and 1.8% had professional degrees with CAIIB.

DATA ANALYSIS

Data were purified using exploratory factor analysis (EFA) and validated through confirmatory factor analysis (CFA).

Scale Purification: Exploratory Factor Analysis

Employer Branding: The EB construct comprised 11 items. The final round of the EFA resulted in two factors: *Economic and Functional Attributes* and *Communication and*

Development Attributes. Kaiser-Meyer-Olkin (KMO) value (0.795) and Bartlett Test of Sphericity (BTS) value (chi-square [χ^2]=2705.639, df=28, and p<0.001) indicated the appropriacy of data for factor analysis. The eigenvalues of both factors are also above the threshold criterion, which represented the significant contribution of these factors (Table 2.1).

Talent Attraction: The final EFA retained seven items under two factors, that is, *Managerial Attraction* and *Organizational Attraction*. The KMO value (0.758) revealed that the sample size is adequate to yield distinct and reliable factors. Furthermore, BTS values (χ^2=1106.563, df=21, and p<0.001) revealed the existence of correlations. The total variance is about 75%. The factor loadings and communality values are above the threshold criterion (Table 2.2).

Talent Identification: TI has emerged into two factors (seven items), that is, *Informal Talent Identification* and *Formal Talent Identification* (Table 2.2). The communality values (CV) and factor loadings (FL) of all the items are above the

TABLE 2.1
Summary of EFA Results of EB

Constructs	EFA							CFA
	M (SD)	KMO	Com.	FL	EV	VE% age	Cronbach's Alpha	SRW
Employer Branding		0.795				85.92		
Communication and Development Attributes(F1)					5.07	43.75	0.95	
Open, flexible, honest communication with employees(EB13)	4.21 (0.87)		0.91	0.92				0.89
Seniors act as coach(EB4)	4.43 (0.89)		0.90	0.92				0.88
Involve employees in decision-making(EB14)	4.11 (0.91)		0.88	0.91				0.87
Recognition of team achievements(EB7)	4.01 (0.92)		0.81	0.88				0.79
Economic and Functional Attributes(F2)					1.81	42.17	0.94	
Activities for societal well-being(EB11)	4.21 (0.73)		0.90	0.93				0.91
Learning in atmosphere(EB3)	4.10 (.87)		0.85	0.89				0.80
Health and safety(EB2)	4.37 (0.84)		0.87	0.88				0.87
Local charities(EB10)	4.10 (0.91)		0.76	0.85				-

TABLE 2.2
Summary of Practice-Wise EFA Results of TM

Practices of Talent Management		KMO	Com.	FL	VE% age	EV	Cronbach's Alpha
Talent Attraction		0.758			75.39		
	Managerial Attraction (F1)				38.19	3.50	0.86
	Devote time and energy (TA7)		0.79	0.88			
	Attract achievement-oriented employees (TA8)		0.79	0.88			
	Attract motivated employees (TA9)		0.73	0.85			
	Organizational Attraction (F2)				37.2	1.78	0.82
	Better career advancement opportunities (TA2)		0.90	0.95			
	Learning and development opportunities (TA3)		0.84	0.92			
	Competitive healthcare benefits (TA5)		0.67	0.69			
	Increases salary linked to individual performance (TA1)		0.56	0.59			
Talent Identification		0.758			76.23		
	Informal Talent Identification (F1)				46.10	3.89	0.91
	Awareness about the level of employees' performance (TI1)		0.90	0.93			
	Feedback from coworkers (TI3)		0.86	0.92			
	Identification on judgment basis (TI5)		0.79	0.87			
	Information in the hand (TI2)		0.61	0.77			
	Formal Talent Identification (F2)				30.13	1.45	0.79
	Nomination process (TI7)		0.84	0.87			

TABLE 2.2
Cont.

Practices of Talent Management		KMO	Com.	FL	VE% age	EV	Cronbach's Alpha
	Segmentation process (TI6)		0.81	0.84			
	Performance appraisal (TI8)		0.53	0.73			
Succession Planning	Higher level management is involved in succession procedure (SP3)	0.788	0.72	0.85	63.55	3.18	0.85
	Update succession plans (SP1)		0.71	0.84			
	Authority is delegated for succession planning (SP4)		0.67	0.81			
	SP is a key part of the decision framework (SP5)		0.58	0.75			
	Use employee data base for SP (SP6)		0.54	0.73			
Talent Development	Identifies development needs (TD8)	.869	0.88	0.94	73.52	5.88	0.95
	Internal development program (TD10)		0.84	0.92			
	Is more accurate and efficient in developing emerging talent (TD7)		0.81	0.90			
	Meet development needs effectively and timely (TD9)		0.80	0.89			
	Develop talent pool for internal recruitment (TD5)		0.79	0.89			
	Offer higher order project assignments to identified employees for their development (TD4)		0.66	0.81			
	Provide opportunities for developing skills and competence (TD6)		0.59	0.77			
	Recognizes managers for development of the talent pool (TD3)		0.52	0.72			

(Continued)

TABLE 2.2
Cont.

Practices of Talent Management		KMO	Com.	FL	VE% age	EV	Cronbach's Alpha
Talent Engagement		0.797			79.38		
	Managerial Level Engagement (F1)				44.17	3.73	0.87
	Opendialogue relationship with employees (TE7)		0.87	0.92			
	Encourage employees for their development (TE6)		0.92	0.91			
	Recognize and praise employees for good work (TE5)		0.87	0.86			
	Organizational Level Engagement (F2)				35.21	1.04	0.76
	Managers are interested in employee well-being (TE3)		0.73	0.85			
	Decision-making authority to do job well (TE4)		0.77	0.82			
	Effectively maintains staffing levels (TE2)		0.60	0.69			
Talent Retention	Retains talented employees who thrive and contribute to growth (TR2)	0.718	0.83	0.91	4.726	7.40	0.92
	Career development programs (TR3)		0.82	0.91			
	Build personal relationship with employees (TR6)		0.75	0.87			
	Provide job security to employees (TR5)		0.62	0.79			
	Management examines the critical factors for retaining the employee (TR1)		0.58	0.76			
	Feedback from employees related to their positions, satisfaction, etc.(TR7)		0.57	0.75			
	Long-term investment plans/ programs for retaining employees (TR4)		0.54	0.74			

Key: Com.=Communality Value, FL=Factor Loading, VE=Variance Explained, EV=Eigenvalue threshold criterion (>0.50). The KMO (0.758) and BTS (χ^2=1106.563, df=21, and $p<0.001$) values support the relevancy of the data for factor analysis. The eigenvalues of both factors are also above the threshold criterion (>1.0), which represented the significant contribution of these factors for designing the TI construct.

Succession Planning: SP has emerged into one factor with five items (64% of the total variance). The KMO value (0.788) and BTS measure (χ^2=655.81, df=10, and p=0.000) indicated the correlation among the items and the appropriateness of the data for EFA. The factor loadings ranged between 0.73 and 0.85 and the communality values ranged between 0.54 and 0.72. Furthermore, the eigenvalue arrived at 3.18, which reflected the relevancy of retained items for SP scale (Table 2.2).

Talent Development: TD has been reduced to eight items under one factor (α= 0.96). The total variance explained has arrived at 73%. Furthermore, the KMO value (0.869) and BTS measure (χ^2=2386.601, df=28, and $p<0.001$) indicated that the data are suitable for conducting factor analysis. All the factor loading and communality values are above the recommended criteria (Table 2.2). The eigenvalue of TD arrived at 5.882.

Talent Engagement: The ten items of this construct got reduced to seven under two factors in the final round of EFA, that is, *Managerial Level Engagement* and *Organizational Level Engagement*. The KMO value arrived at 0.797 (χ^2= 655.959, df=15, and $p<0.001$) and represents significant correlation. The total variance explained about 79% of the factor loading (Table 2.2).

Talent Retention: Initially, EFA was carried out on the ten items, which got reduced to seven items. The KMO value (0.718) and BTS (χ^2=1937.826, df=21, and $p<0.001$) indicated inter-item correlation and adequacy of the data. The factor loadings of seven items ranged from 0.74 to 0.91, with an acceptable range of communalities (>0.50). The variance explained by TR is about 67% (Table 2.2). The Cronbach's alpha value of 0.92 shows high internal consistency.

Overall Talent Management Practices in Public Banks: After conducting the EFA on each practice individually, the EFA was again carried on all the practices collectively with retained items in the last step. As a result, 40items got reduced to 20 and converged under six factors (Table 2.3), namely, talent development (F1), succession planning (F2), talent retention (F3), talent attraction (F4), informal talent identification (F5) and talent engagement (F6). Furthermore, the KMO value (0.735) and BTS (χ^2=4113.14, df=190, and $p<0.001$) indicated the presence of correlation among the variables and appropriacy of data for the application of factor analysis. The six-factor solution explained about 76% of the total variance. The factor loadings and communality values are also in the acceptable range. Besides this, the internal consistency reliability estimates (>0.70) for the six practices of TM construct.

RESULTS

SCALE VALIDATION: CONFIRMATORY FACTOR ANALYSIS

A CFA was run to analyze the measurement properties of the various scales, which provides an assessment of reliability and validity of the scales.

Employer Branding: The goodness of model fit indices resulted in good model fit (see Table 2.4). The results also indicated that communication and development

TABLE 2.3
Summary of Factor Analysis of Talent Management

Construct	EFA							CFA
	M (SD)	KMO	Com.	FL	VE% age	EV	Cronbach's Alpha	SRW
Talent Management		0.735			76.47			
Talent Development (F1)					19.27	6.45	0.95	
Identifies development needs (TD8)	4.11 (0.88)		0.90	0.91				0.96
Meet development needs effectively and timely (TD9)	4.01 (1.01)		0.88	0.90				0.85
Is more accurate and efficient in developing emerging talent (TD7)	4.07 (0.90)		0.85	0.89				0.92
Internal development program(TD10)	4.07 (0.95)		0.90	0.89				0.88
Succession Planning (F2)					14.20	2.43	0.84	
Update succession plans (SP1)	4.00 (1.07)		0.76	0.84				-
Higher level management is involved in succession procedure (SP3)	4.03 (0.96)		0.77	0.77				0.66
Use employee data base for SP (SP6)	4.00 (1.13)		0.73	0.75				0.89
SP is a key part of the decision (SP5)	4.08 (0.95)		0.68	0.74				0.74
Talent Retention (F3)					13.26	1.95	0.85	
Provide job security to employees (TR5)	4.12 (0.86)		0.81	0.88				0.87
Management examines the critical factors for retaining the employee (TR1)	4.04 (0.81)		0.75	0.83				0.80
Feedback from employees related to their positions, satisfaction, etc. (TR7)	4.13 (0.79)		0.70	0.74				0.76
Talent Attraction (F4)					11.03	1.86	0.84	
Learning and development opportunities (TA3)	4.09 (0.93)		0.90	0.93				0.56
Better career advancement opportunities (TA2)	4.01 (0.99)		0.91	0.91				0.59
Competitive healthcare benefits(TA5)	4.29 (0.72)		0.73	0.55				0.91
Informal Talent Identification (F5)					10.70	1.43	0.79	
Feedback from co-workers (TI3)	4.11 (0.98)		0.91	0.92				0.92

TABLE 2.3
Cont.

Construct	EFA							CFA
	M (SD)	KMO	Com.	FL	VE% age	EV	Cronbach's Alpha	SRW
Information in the hand (TI2)	4.08 (0.89)		0.89	0.83				0.81
Identification on judgment basis (TI5)	3.88 (1.18)		0.61	0.68				-
Talent Engagement (F6)					8.01	1.21	0.76	
Effectively maintains staffing levels (TE2)	4.37 (0.53)		0.65	0.79				0.55
Managers are interested in employee well-being (TE3)	4.38 (0.54)		0.55	0.70				0.78
Decision-making authority to do job well (TE4)	4.29 (0.60)		0.50	0.65				0.51

Key: M=Mean, SD=Standard Deviation, Com. =Communality Value, FL=Factor Loading, VE=Variance Explained, EV=Eigenvalue

TABLE 2.4
Reliability, Validity and Goodness of Model Fit Indices

Construct	Cronbach's Alpha	CR	AVE	χ^2/df	GFI	AGFI	NFI	CFI	RMSEA
Employer Branding	0.82	0.96	0.87	3.818	0.955	0.895	0.978	0.984	0.080
TM (first-order model)	-	-	-	2.675	0.890	0.838	0.903	0.939	0.078
TM (second-order model)	0.92	0.99	0.99	3.040	0.865	0.813	0.882	0.916	.0087

attributes had higher Standardized Regression Weights (SRW) values as compared to economic and functional attributes.

Talent Management Practices: CFA was run on the six-factor model of talent management, which was extracted after EFA. In this analysis, a two-step procedure was adopted, that is, the first-order CFA model and the second-order CFA model. In the first-order factor model, 20 items got reduced to 18 items. The fit indices of the first-order model provided a good fit to the data (see Table 2.4). The results of the second-order factor model with six first-layer latent constructs also revealed good fit indices (Table 2.4). Therefore, this study used the second-order factor model of talent management to examine reliability and validity in the public banks.

TABLE 2.5
Discriminant Validity and Correlation Analysis

Constructs	TM	EB
TM	0.99	
EB	0.47**	0.93

Note: Values on the diagonal axis represent the square root of AVE. Values off the diagonal are correlation values. ** $p<0.01$.

Reliability

Cronbach's Alpha: This study computed that the Cronbach's alpha (α) values for EB and TM are above the recommended criterion of 0.70, and thus the scales are considered to be reliable (Table 2.4).

Composite Reliability: The study found the composite reliability of the scale items to be above the critical level of 0.70 (Table 2.4), which suggest a high degree of reliability of the constructs.

Validity

Content Validity: Content validity of the scales used in the current research was established by extracting relevant items from the extant literature and developing self-designed items through deliberations with the subject experts. These items were then discussed with academicians and bank managers.

Convergent Validity: The results of this study indicated that the average variance extracted (AVE) values of overall talent management and EB are above the threshold criterion of .50, which yield the existence of convergent validity (Table 2.4).

Discriminant Validity: The results revealed that the correlation between measurement scales estimates is not greater than the square root of AVE, which proved discriminant validity in the study (Table 2.5).

Nomological Validity: This study has examined nomological validity by testing the relationship between EB and organizational performance. Aldousari et al. (2017) stated a positive relationship between EB and organizational performance. The result of the study revealed that EB significantly effects organizational performance [χ^2/df=3.504, root mean square residual (RMR)=0.050, goodness-of-fit index (GFI)=0.873, adjusted goodness-of-fit index (AGFI)=0.778, normed fit index (NFI)=0.821, comparative fit index (CFI)=0.864, and root-mean-square error of approximation (RMSEA)=0.081], which confirmed the nomological validity of the EB scale.

Common Method Variance

Since bank employees were contacted for collecting the data, this can create problems of biasness (Podsakoff et al., 2003). To resolve this problem, Harman's single-factor test method, which is most commonly used by researchers, was used to examine the

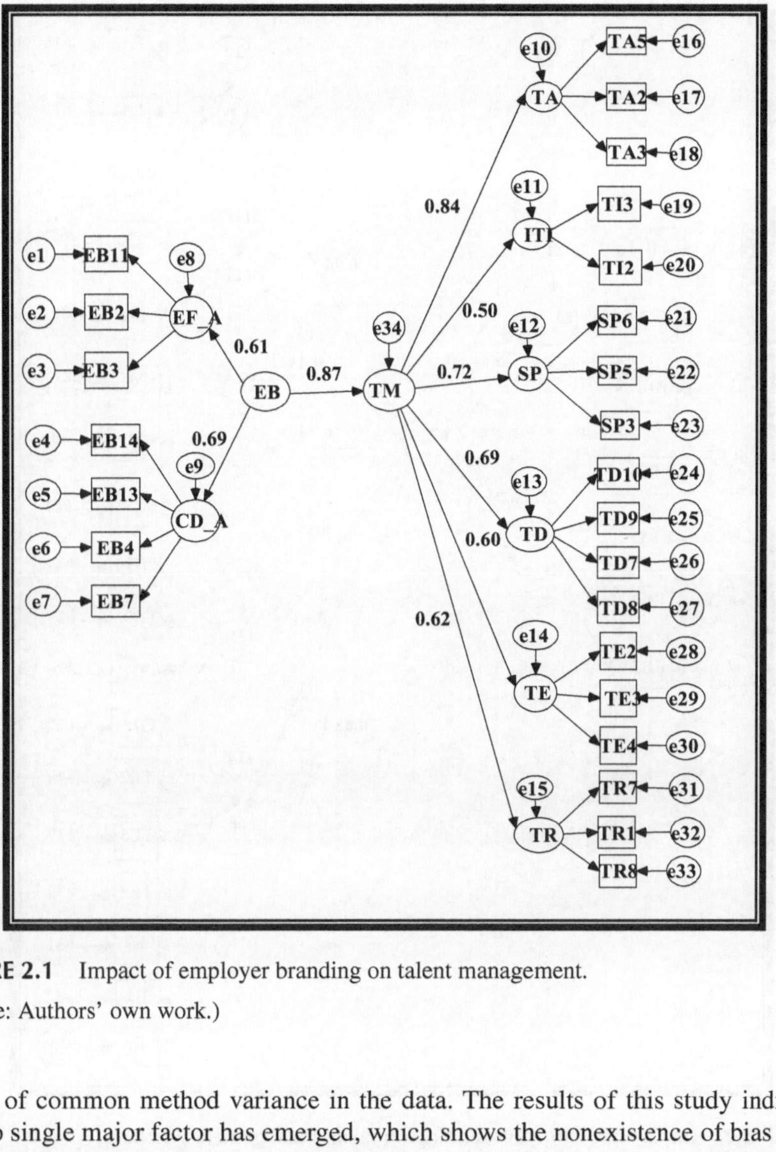

FIGURE 2.1 Impact of employer branding on talent management.
(Source: Authors' own work.)

extent of common method variance in the data. The results of this study indicated that no single major factor has emerged, which shows the nonexistence of bias in the data. Furthermore, CFA also helps to minimize the problem of biasness (Richardson et al., 2009).

Structural Equation Modeling (SEM)

This study used SEM for testing the hypothesized relationships (Arbuckle & Wothke, 2004). In the present study, the relationship between EB and talent management has been assessed. The result revealed that EB has a significant and positive impact on talent management (EB®TM=0.87***, Figure 2.1). Therefore, the first hypothesis (i.e., *employer branding significantly affects talent management*) is confirmed.

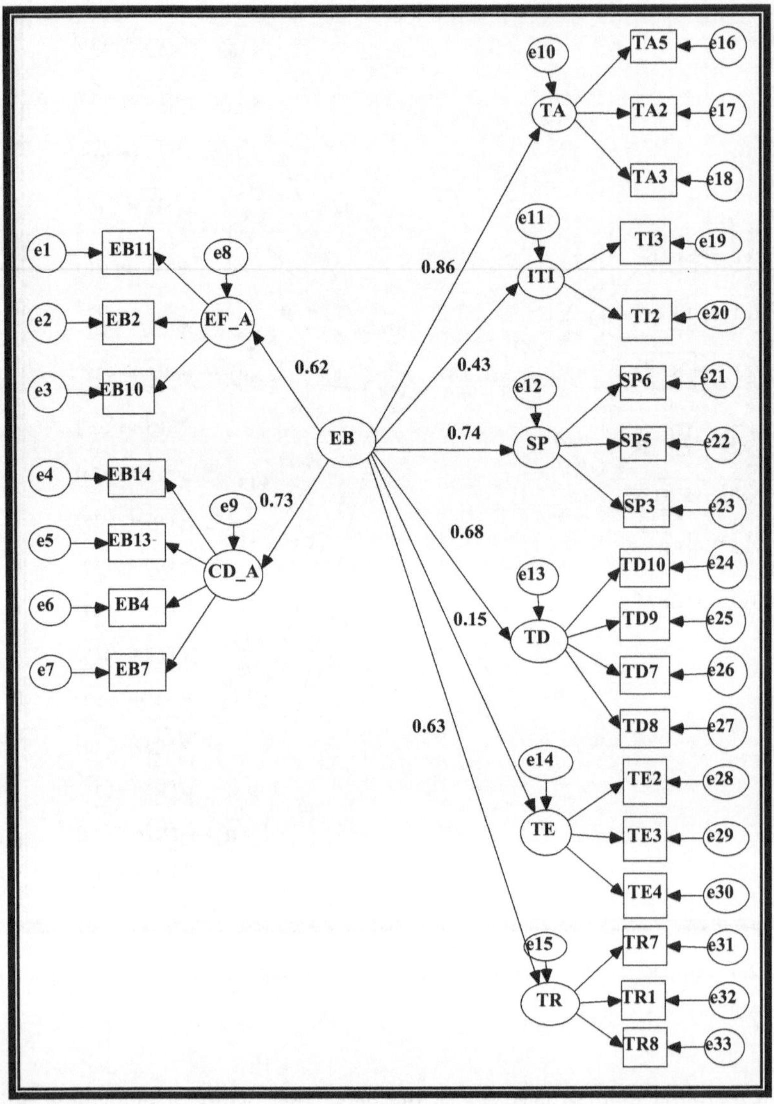

FIGURE 2.2 Impact of employer branding on talent management practices.
(Source: Authors' own work.)

The model has average fit (χ^2/df=4.730, RMR=0.045, GFI=0.801, AGFI=0.758, NFI=0.860, CFI=0.922, RMSEA=0.088). Furthermore, a new model also has been designed where the relationship between EB and talent management practices has been assessed. The results revealed that EB has significant impact on all the talent management practices (EB→TA=0.86***; EB→ITI=0.43***; EB→SP=0.74***; EB→TD=0.68***, EB→TE=0.15*; and EB→TR=0.63***, Figure 2.2). Further, the model has average fit ($\chi 2/df=4.729$, RMR=0.046, GFI=0.872, AGFI=0.816, NFI=0.906, CFI=0.939, RMSEA=0.071). Hence, hypotheses 1(a), 1(b), 1(c), 1(d), 1(e), and 1(f) (i.e., *employer branding significantly affects talent management practices*) are supported. As per this model (Figure 2), EB has the highest impact on talent attraction.

DISCUSSION

In today's modern HRM, EB is very crucial for implementing talent management in the organization. According to this study, public banks have good reputation (mean [M]=4.35). This study validates the EB scale, which consisted of two factors, that is, economic and functional attributes and communication and developmental attributes. Public banks adopt and practice both employee- and society-related activities to develop their bank's image as good employers (M=4.37). These banks provide comfortable work environment, equitable wages, and justice to their employees. Furthermore, banks also have open, flexible, and honest communication process with their employees (M=4.51). Banks also contribute toward the personal and career development of their employees (M=4.49), which help to develop and maintain the good image of the organization. EB plays an important role as a predictor for dealing with TM and all its practices, that is, talent attraction, informal talent identification, succession planning, talent development, talent engagement, and talent retention in the banks. The presence of effective EB strategy in the organization helps to builds the "value proposition" image of the organization as a good place to work, which helps in attracting, identifying, positioning, developing, engaging, and retaining a high-potential workforce successfully. Moreover, the results of this study also revealed that talent management is being practiced in public banks (M=4.29).

MANAGERIAL IMPLICATIONS

The study results also found that EB has a strong and significant impact on TM and its practices. It signifies that EB is an essential component of effective and proper implementation of TM. As poor EB can lead to low talent attraction, low employee development, low work engagement of the talent pool, and high turnover rate, therefore the following strategic implications are recommended for bank managers to enhancing the EB in the respective sector:

- Managers should give special attention to the following attributes of employer branding: economic, functional, communication, and development attributes for managing high-potential employees from hiring to retirement period.
- Under economic and functional attributes, banks should focus on attractive compensation package, recognition/appreciation from management, health and safety work environment, hands-on interdepartmental experience, work-related policies and flexi time, promotion opportunities on the basis of talent, and workplace flexibility for high-potential employees.

- Under communication and development elements, managers should develop a culture of sharing and continuous improvement, opportunities to learn new skills, internal career opportunities, opportunity to share ideas, and springboard for future employment, and treat each employee with respect and integrity.
- Banks should also give more attention to develop social values, such as good relationship with superiors and colleagues, and should support and encourage colleagues for developing a good image. Moreover, novel work practices/forward-thinking also helps to build good image. It attracts and retains high-potential employees.

LIMITATIONS OF THE STUDY AND FUTURE RESEARCH DIRECTION

Following are the limitations of this study: first, this study has a single source of respondents, namely, bank employees, who might have been guided by their likes and dislikes. Second, this study is cross-sectional in nature, which can affect the extent of causality of relationship.

In the future, a longitudinal study should be conducted. Furthermore, researchers should test the mediating role of talent management between EB and other variables such as organizational commitment, employee performance, job satisfaction, and trust in leadership. Researchers should also try to find out the mediators between EB and talent management.

CONCLUSION

This chapter suggests that the clearest link between EB and talent management practices. EB helps to attract, identify, develop, engage and retain high potential employees in the organizations. The combination of EB and talent management practices can make the organization for both existing and potential employees.

REFERENCES

Anderson, J.C., & Gerbing, D.W. (1988). Structural equation modeling in practice: A review and recommended two-step approach. *Psychological Bulletin, 103*(3), 411–423.

Arbuckle, J., & Wothke, W. (2004). *Amos 4 users guide*. Chicago, IL: Small Waters Corporation.

Bano, S., Khan, M.A., Rehman, Q.H.U., & Humayoun, A.A. (2010).Schematizing talent management: A core business issue. *Far East Journal of Psychology and Business, 2*(1), 416.

Bhatnagar, J. (2007). Talent management strategy of employee engagement in Indian ITES employees: Key to retention. *Employee Relations, 29*(6), 640663.

Botha, A., Bussin, M., & De Swardt, L. (2011). An employer brand predictive model for talent attraction and retention. *SA Journal of Human Resource Management, 9*(1), 112.

Collings, D.G., & Mellahi, K. (2009). Strategic talent management: A review and research agenda. *Human Resource Management Review, 19*(4), 304–313.

Das, T.V., & Rao, H.P. (2012). Employer brand in India: A strategic HR tool for competitive advantage. *Advances in Management, 5*(1), 23–27.

Davies, G. (2008). Employer branding and its influence on managers. *European Journal of Marketing, 42* (5/6), 667–681.

Dawn, S.K., & Biswas, S. (2011). Employer branding: A new strategic dimension of Indian corporations. *Asian Journal of Management Research, 1*(1), 21–33.

Edwards, M.R., &Bartlett, T.E. (1983). Innovations in talent identification. *SAM Advanced Management Journal, 48*(4), 16–22.

Edwards, M.R. (2010). An integrative review of employer branding and OB theory. *Personnel Review, 39*(1), 5–23.

Highhouse, S., Lievens, F., & Sinar, E.F. (2003). Measuring attraction to organisations. *Educational and Psychological Measurement, 63*(6), 986–1001.

Horvathova, P., & Durdova, I. (2010). The level of talent management usage at human resources management in organizations of the Moravian-Silesian region. *Business and Economic Horizons, 3*(3), 58–67.

Ibrahim, N.S., Hashim, J., & Rahman, R.A. (2018). The impact of employer branding and career growth on talent retention: The mediating role of recruitment practices in the Malaysian public sector. *International Journal of Academic Research in Business and Social Sciences, 8*(6), 1034–1039.

Jyoti, J., & Rani, R. (2014). Exploring talent management practices: Antecedents and consequences. *International Journal of Management Concepts and Philosophy, 8*(4), 220–248.

Kapoor, V. (2010). Employer branding: A study of its relevance in India. *The IUP Journal of Brand Management, 7*(1), 51–75.

Lockwood, N.R. (2007). Leveraging employee engagement for competitive advantage: HR's strategic role. *The Society for Human Resource Management Research Quarterly, 52*(3), 2–11.

Lockwood, N.R. (2010). Employer brand in India: A strategic HR tool. *The Society for Human Resource Management Research Quarterly*, 2–13.

Mahesh, R., & Suresh, B.H. (2019). Employer branding as an HR tool for talent management – An overview. *International Journal of Management Studies, 6*(1/5), 74–80.

Mandhanya, Y., & Shah, M. (2010). Employer branding:A tool for talent management. *Global Management Review, 4*(2), 43–48.

Mihalcea, A. (2017). Employer branding and talent management in the digital age. *Management Dynamics in the Knowledge Economy, 5*(2), 289–306.

Moroko, L., & Uncles, M.D. (2008). Characteristics of successful employer brands. *Brand Management, 16*(3), 160–175.

Piansoongnern, O., Anurit, P., & Kuiyawattananonta, S. (2011). Talent management in Thai cement companies: A study of strategies and factors influencing employee engagement. *African Journal of Business Management, 5*(5), 1578–1583.

Podsakoff, P.M., MacKenzie, S.B., Lee, J.Y., & Podsakoff, N.P. (2003). Common method biases in behavioral research: A critical review of the literature and recommended remedies. *Journal of Applied Psychology, 88*(3), 879–903.

Priyadarshi, P. (2011). Employer brand image as predictor of employee satisfaction, affective commitment and turnover. *The Indian Journal of Industrial Relations, 46*(3), 510–522.

Rana, G., & Sharma, R. (2019). Assessing impact of employer branding on job engagement: A study of banking sector. *Emerging Economy Studies, 5*(1), 7–21. https://doi.org/10.1177/2394901519825543.

Rana, G., Sharma, R., Singh, S., & Jain, V. (2019). Impact of employer branding on job engagement and organizational commitment in Indian IT sector. *International Journal of Risk and Contingency Management (IJRCM), 8*(3), 1–17. doi:10.4018/IJRCM.2019070101

Richardson, H.A., Simmering, M.J., & Sturman, M.C. (2009). A tale of three perspectives: Examining post hoc statistical techniques for detection and correction of common method variance. *Organizational Research Methods, 12*(4), 762–800.

Ritson, M. (2002). Marketing and HE collaborate to harness employer brand power. *Marketing*, 24. Available at: www.campaignlive.co.uk/article/opinion-marketing-hr-collaborate-harness-employer-brandpower/162068?src_site=marketingmagazine (accessed 27 January 2019).

Sharma, R., Jain, V., & Singh, S.P. (2018). The impact of employer branding on organizational commitment in Indian IT sector. *IOSR Journal of Business and Management,*20(1), 49–54. https://doi.org/10.9790/487X-2001054954

Srivastava, P., & Bhatnagar, J. (2010). Employer brand for talent acquisition: An exploration towards its measurement. *The Journal of Business Perspective, 14*(1/2), 25–34.

Vaijayanthi, P., Roy, R., Shreenivasan, A., & Srivathsan, J. (2011). Employer branding as an antecedent to organisation commitment: An empirical study. *International Journal of Global Business, 4*(2), 91–106.

Wilden, R.M., Gudergan, S., &Lings, I.N. (2010). Employer branding: Strategic implications for staff recruitment. *Journal of Marketing Management, 26*(1–2), 56–73.

Yarnall, J. (2011). Maximizing the effectiveness of talent pools: A review of case study literature. *Leadership and Organization Development Journal, 32*(5), 510–526.

3 The Role of Employer Branding in the Creation of Powerful Corporate Brands

Harsh Mishra and Aditi Sharma

INTRODUCTION

The idea of transforming the corporate into a brand is and has been present in the branding literature since 1950 (Balmer, 1995). In fact, it can be safely asserted that almost all of the world's original brands were corporate brands as most of the companies produced only one product in the beginning which became synonymous with the company. The second half of the 20th century witnessed rapid advancement in technology and the advent of globalization in its modern form. This led to a single company offering multiple products across multiple countries. This period also saw the emergence of developing economies across the world and numerous companies offering similar products in similar markets. With product differentiation becoming more and more of a concept than a reality, the companies were compelled to brand their products in order to remain competitive in their respective markets. However, with ever increasing number of similar products available in the market, it became incrementally more difficult and expensive for brands to retain their competitive advantage and market share using product branding. This led people interested in profitability and sustainability of a company to look at other ways of establishing a brand in order to remain relevant in an increasingly competitive market space. Kennedy (1977) argued that the companies need to look internally and consciously create a "company image" reflecting organizational, cultural, and brand values. Although, she did not explicitly mention corporate brand, all the attributes of the "company image" she mentioned may be found in modern branding literature. Olins (1978) proposed the idea of monolithic, endorsed, and branded identities. His concept of monolithic identities was similar to the modern idea of corporate brands. Although, the concept of corporate brand was in existence in one form or another for quite some time, it started to gain ascendancy in branding literature during the last decade of the 20th century and since then many scholars have attempted to delineate different aspects of corporate branding (Abimbola, Trueman, Iglesias, Abratt & Kleyn, 2012; Balmer, 1995, 2001; Balmer & Gray, 2003; Biraghi & Gambetti, 2015; Hatch & Schultz, 2003; Hatch & Schultz, 2001; King, 1991; Knox & Bickerton, 2003; Mitchell, 1997; Schultz & de Chernatony, 2002). The importance of creating and augmenting corporate brands becomes conspicuous when we take cognizance of the fact that world's top 20 brands are primarily corporate brands comprising industry leaders such as Apple, Google, Amazon, Microsoft, BMW, Samsung, Disney, etc. (Best Global Brands 2019 Rankings, 2019). Companies such as Procter & Gamble, which have been strong proponents of product brands, having built huge multinational conglomerates riding on the success of those brands, have also later realized the importance of corporate branding (Hatch & Schultz, 2001). The fact that so many of the top brands in the world are corporate brands renders it imperative for most organizations to look at some form of corporate branding in order to remain competitive and sustainable in a global market place which is becoming more cut throat with each passing day.

The primary distinction between a corporate brand and a product brand is that the former focuses on all the stakeholders while the latter focuses primarily on customers (Balmer & Gray, 2003). Kennedy (1977) drew the attention of the branding community towards the importance of employees in building a propitious company image. King (1991, p. 8) proposed the idea of "staff as brand-builders." Since then several scholars have extolled the role of employees in the process of corporate branding (Balmer & Gray, 2003; Harris & de Chernatony, 2001; Hatch & Shcultz, 2001; Potgieter & Doubell, 2020; Punjaisri & Wilson, 2017). Firstly, two out of the three "strategic stars" required for the development of a successful corporate brand, viz., vision, culture, and image, as proposed by Hatch & Schultz (2001), are fully dependent upon the employees of the organization. It is the top management that determines the vision of a corporation while it is the employees who are responsible for translating that vision into reality. Also, the culture of the organization as reflected in its values, ethics, and attitude towards its various stakeholders is communicated primarily through the employees, be it the top management or the frontline salesperson. Therefore, it becomes imperative for the corporations to recruit the appropriate set of employees whose personalities and values are in consonance with the values and the culture of the brand because only then they may be able to become the flag bearers of the corporate brand. This is where the role of employer branding becomes important. A successful employer branding strategy may help the corporations in recruiting and retaining the employees who identify with the brand values of the organization and therefore, become natural promoters of the corporate brand.

THE CORPORATE BRAND

Evolution of humongous multinational corporations, in an increasingly globalized and competitive environment, has rendered it essential for organizations to look upon themselves as brands. Projection of corporations as unified corporate brands has become vital for their functioning in a globalized environment. The focus has gradually started to shift from product brands to corporate brands (Aaker, 1996; Balmer, 1995; Dowling, 2001; Hatch & Schultz, 2001, 2003; Olins, 2000) and employees are at the forefront of this change due the role they play in the creation of corporate brands. Hatch and Schultz (2001) suggested that the concept of product differentiation is becoming more and more untenable owing to rapid globalization and disintegrating marketplaces. Under such circumstances, corporations which are able to integrate the vision formulated by the top management with the overall organizational culture and image may find themselves in a better position to the changing realities of the dynamic business environment than those who are dependent solely on product branding. This is not to suggest that product brands have been relegated to an inferior position vis-à-vis corporate brands. In fact, many corporate brands such as Virgin have originated from product brands. The product brands are still successful in spheres of homogenized mass consumption while the emergence of corporate brands may be attributed to the growing need of companies to cater to pluralistic, multicultural, and multiethnic societies (Palazzo & Basu, 2007). One reason for this shift in focus is that stakeholders – external as well as internal – are becoming more and more inquisitive and demanding. They are now showing active interest in not only the "ends" but the "means" as well. Many top multinational companies, including some of the biggest brands such as Nestle, McDonald, Johnson & Johnson, etc., have been forced to mend their ways due to continuous pressure from the discerning consumers who were not only interested in the products they were buying but were also concerned about the ethical, environmental, and production policies of the companies making those products (Kane, 2015). Increased opportunities of communication that the stakeholders have been endowed with owing to rapid expansion of the Internet – especially social media – since the turn of the century have made it almost impossible for the companies to take refuge behind their product brands. Balmer (1995) asserted that the stakeholders are becoming more perceptive and inquisitive about the companies behind the products they buy and it would become increasingly difficult for corporations to conceal their functioning especially in terms of their ethical and environmental policies. This cannot be truer in the modern era of communicative expansion where a company's market share may vanish overnight just because a video depicting their unethical employment practices has gone viral on a social media platform. Corporations have started to realize that stakeholders associate certain values – positive or negative – with them. This realization has led conscious efforts on the part of these corporations to manage their corporate brand actively (Christensen, Morsing & Cheney, 2008). However, the task of managing corporate brands is much more complex than the task of managing product brands primarily because ubiquitous nature of the corporate brands and also because they have to focus on all the stakeholders instead of external stakeholders only. The positive or negative values associated with a corporate brand get extrapolated upon all the product brands associated with it while the reverse of this is not usually true. For instance, the recent controversies surrounding Nestlé's product brand Maggi have not made that much of an impact upon other brands in its repertoire. But if the controversy was associated with the corporate brand Nestle, then the repercussions would have been experienced by all other brands associated with it. Corporate branding, unlike product branding which can be handled by one or two departments in the organization, requires involvement of all the departments of the organization. It requires an integration of internal as well as external communications in order to achieve its objective of portraying a unified image of the organization. Also, developing and sustaining strong corporate brands requires integration of almost all the functions of an organization ranging from marketing to human resources.

Corporate branding comprises corporate identity, corporate image, and corporate reputation. All the stakeholders have some role to play in the process of creating corporate brands. The basic premise behind corporate branding is that the customers are becoming increasingly sophisticated showing greater propensity to "buy" the company behind the product than the product itself. For instance, people with more patriotic inclinations would be more likely to a product made in their own country even if the substitute made by a foreign country is more technologically advanced or less expensive. The "companies" that the customers buy – or do not buy– are made up of diverse components ranging from the products they sell to the image they project. Most of these components are dependent upon quality of human resources they have. Human resource, although a very important component of successful brands, has been quite often neglected from the perspective of corporate branding. Corporations generally have an inclination towards consumers while defining their brand values and identity, while neglecting the employees who have to actually carry forward that brand value (Brønn, 2002). Employees are one of the most important factors from the point of creating and managing successful corporate brands. They are the face of any organization. Foster, Punjaisri, and Cheng (2010) – while attempting to explore the relationship among corporate, employer, and internal branding – have proposed that employer branding, along with internal branding, plays a crucial part in the creation of effective and powerful corporate brands. The employees are important because they carry the brand image of their organization with them round the clock. As members of the society, they continuously remain in touch with the existing as well as prospective customers and other stakeholders of the organization they work for. Corporate brand image is dependent upon the behavior of employees especially in case of service brands where employees are at the forefront of customer dealings. If the employees engage with customers in a cordial manner, be attentive to their needs, are empowered to take action to fulfil their demands, are quick to respond to their concerns and are skilled, then it would manifest positively in customer perceptions and attitudes towards the corporate brand and they are most likely to become loyal to the brand in the long run (Aaker, 2004). However, finding such employees and transforming them from mere service or skill providers to brand ambassadors is not an easy task. Employer branding has emerged as a possible solution to this problem.

INTEGRATING EMPLOYER BRANDING WITH CORPORATE BRANDING

During past few decades, employer branding has emerged as an essential component of modern multinational conglomerates' business strategy. The concept of "employer brand" was introduced by Tim Ambler and Simon Barrow in 1996. They proposed that the benefits offered by an employer brand to its employees are quite similar to those which are apparently offered by a traditional product brand to its customers. Employee development opportunities, monetary rewards, and close affinity with the organization are some of the major components of Employer Branding (Ambler and Barrow, 1996). Ind (1998) proposed that the process of recruitment should be considered as an exercise in branding. It involves applying the marketing principles to human resources in the organization to develop a distinct reputation among its current and prospective employees, and to attain competitive advantage by hiring and retaining a talented pool of people. Employer branding as a differentiating strategy implies that an employer makes conscious efforts to portray itself distinct from other players in the labor market (Gehrels, 2019). Biswas and Suar (2016) carried out a research study in India to find out the consequences of employer branding for a corporate brand. They found that employer branding is influenced by organizational support system for employees, truthful information about job roles, perceived prestige of the organization, fair reward system, trust in the organization, leadership from top management and the manner in which a corporation fulfils its social obligations. An overall positive perception with respect to the aforementioned factors leads to better performance by the corporation concerned in terms of financial and nonfinancial parameters.

As the global "war for talent" (Gehrels, 2019) is becoming more and more intense on account of talent shortages, the top companies are applying the concept of employer branding on similar lines of corporate branding (Mosley, 2015). Their focus has been on alluring and retaining the best employees who can be trained to imbibe the fundamental values of the organization so as to transform them into brand ambassadors of the corporate brand. In the times of rigorous competition, volatilities, and uncertainties in the business world, there is a pressing need for hiring the right kind of people who can deliver results and contribute to the growth of organization. Attracting and retaining the manpower has always been one of the key functions of the Human Resource Management (HRM) department. The ascendance of corporate branding during the past three decades has added another responsibility to the HRM department. Now, they need to ensure "person-organization fit (P-O)" (Kristof, 1996, as cited in Potgieter & Doubell, 2020, p. 113) with respect to the values, the culture, and the image of the corporation while hiring employees. During the past two decades, Employer branding has evolved as a critical strategy for attracting a talented pool of employees to the organization. Employer branding has been actively deployed by multinational corporations such as Google, Facebook, Marriot, etc. for attaining competitive advantage. It has enabled them to project themselves as "workplace-of-choice" to the existing as well as the prospective employees. Besides the usual advantages related to the acquisition of "world class employees," it has offered the organization other benefits including higher involvement, improved motivational levels, and better commitment from the employees. Thus, employer branding may be leveraged to strengthen the corporate brands.

Employer branding became an important subject matter of HRM literature due to the rising popularity of the concept of branding in the marketing arena; the concept of employee engagement catching the attention of the academicians and the Human Resource (HR) managers; talent becoming a scarce commodity especially in case of knowledge-based workers and the realization that there is a significant correlation between the HRM practices adopted by an organization and its growth (CIPD, 2007; as cited in Biswas and Suar, 2016, p. 58). The challenge before the organizations is that talented manpower always has a preferential choice regarding the organization where they would like to work and attracting them becomes a herculean task (Tanwar & Kumar, 2019). Employer branding is based on the principles of effective communication with an idea of attracting and retaining the better among the best people available in the competitive labor market (Backhaus & Tikoo, 2004; Turban & Greening, 1996). In a pluralistic society and a post-industrialized working environment where the demand for knowledge workers is increasing rapidly, effective implementation of employer branding practices can create a unique value proposition for an organization's current and future employees, catapulting the organization into the league of most preferred employers (Edlinger, 2015). Such employers are usually able to hire the employees whose values and culture are in consonance with the organization's values and culture. Recruitment of such employees helps the organization in pushing forward its corporate branding agenda.

Employer branding is a strategy employed by the organizations to attract the best of talented candidates available externally as well as manage their highly skilled employees internally. Employer branding is aimed at managing the perceptions of current employees, prospective employees, and relevant stakeholders so that they may actively contribute in the management of the corporate brand. Employer branding helps an organization in creating an image of desirability in the minds of prospective and current employees. The employee is convinced that the search for the kind of organization she is scouting for culminates here. Organizations participate in the best employer surveys in their endeavor to endear the best-suited employees in the market (Saini, Rai & Chaudhary, 2014). A survey by Future Today conducted on 3773 students studying in top Moscow Universities revealed two categories of attractive employers: the "WOW companies" and the "strategist." The former attracted the attention of young employees on the basis of a strong corporate brand image created in their minds through their distinct products while the latter focused upon building relationship with youngsters (Kucherov & Zamulin, 2016). A job aspirant is more likely to apply for a job where she knows about the brand either due to its products or amazing customer service (Collins, 2007). Such job aspirants are quite valuable from the point of view of corporate branding as they understand the brand and already have a positive attitude towards it.

At initial levels of employer branding, the company would have to bear the costs of building a brand but at later stages, it would help the organization in developing a stable and symbiotic relationship with its employees (Kucherov & Zamulin, 2016) as well as "future-proof" its corporate reputation (Martin, Gollan & Grigg, 2011). This stable and symbiotic relationship would lead to the creation of a workforce that understands the brand and the values that it carries. Thus, employer branding may also be understood as a mutual agreement between a corporation and its employees with respect to the reasons behind choosing and continuing the job over a period of time. The idea is to present the organization as a brand, which is different from the run of the mill employers, with fascinating values and culture pertinent to the individual's aspirations and expectations, besides ensuring that these attributes of the brand do not remain only a promise and are delivered to the employees continuously (Biswas, 2013). Employees prefer to have a clear communication about the culture, the values, and the vision of the organization so that they may also have a clear idea about what is expected of them and the manner in which they may fulfil their obligations to the organization.

EMPLOYER BRANDING, PSYCHOLOGICAL CONTRACT, AND PSYCHOLOGICAL OWNERSHIP

When employees join an organization, they formulate certain perceptions about the organization that are based on their understanding of the organization's expectations from them and also their expectations from the organization. The employees perceive that the organization expects them to be loyal and committed to the organization goals while in return they expect the organization to take care of their needs (Miles, and Mangold, 2004). The perceptual process shapes the association between the employee and the organization which implies that a positive perception would lead to positive outcomes while a negative perception may have damaging consequences for both the parties (Biswas & Suar, 2016). Employer branding as a process strives to create a positive image about the organization in the mind of the employee. The employees' perception of the employer brand shapes their attitude and behavior towards the organization and its customers. The perception of the fulfilment of a psychological contract between the employee and the organization influences the process of brand building in the minds of the employee (Xiong, Kinga & Piehler, 2013). Psychological contract motivates employees to portray a positive organizational image before the customers through their actions (Gelb & Rangarajan, 2014) and plays a pivotal role in internalization of organizational values (Miles & Mangold, 2004).

Psychological ownership is a process in which an employee identifies with the brand and has a feeling of ownership towards the corporate brand and adopts an altruistic approach towards brand building activities. The brand's psychological ownership leads to a positive attitude and proactive behavior from the employee and leads to abandonment of self-gains (Chang, Chiang & Han, 2012). When the employees understand their role in creating a brand, their sense of psychological ownership towards the organization's brand is enhanced resulting in pro-brand attitudes and behaviors (Xiong et al., 2013). Yakimova, Mavondo, Freeman, and Stuart (2017) suggested that employees who have strong affinity with core brand values and identify closely with the vision of the brand exhibit "brand championship" behavior and consequently, help in the process of corporate branding. Ind (1998) also argued that when the employees understand and believe in the values of the brand, they perform much better and actively contribute to the promotion of the corporate brand.

ROLE OF EMPLOYEES IN CORPORATE BRANDING

Corporate branding literature reiterates that successful corporate branding is dependent upon alignment of vision and culture of the organization with the experiences that the stakeholders have with the brand (Balmer, 1995; Harris & de Chernatony, 2001; Hatch & Schultz, 2001, 2003; Ind, 1998). The aforementioned integration may be possible only with the active participation of the employees at every level of the organization in the process of branding. Potgieter and Doubell (2020) found that employer branding and corporate branding have a significant relationship. The study also found out that the stakeholders' perception of the organization and the manner in which various stakeholders engage with an organization depend on the behavior and conduct of the employees of the organization. Previous studies have also revealed that employees are the most important link between an organization and its stakeholders (Balmer & Soenen, 1999; Balmer & Gray, 2003; Hatch & Schultz, 2001; Harris & de Chernatony, 2001; Hoppe, 2018; Wilson, 2001; Yakimova, Mavondo, Freeman, & Stuart, 2017). These findings make it conspicuous that creation of a powerful corporate brand requires discovering and employing people whose personality, ethics, and values are in consonance with the corporation's culture and brand values. The more the employees identify with the corporate culture and the corporate values, more will be the internalization of the brand, leading to the employees becoming more active participants in the corporate branding process. Hatch and Schultz (2001) argued that for corporate branding to be successful values proclaimed by a corporate brand must be in sync with the connotations that members of the organization derive from it and employ while dispensing their duties. If a corporation's claimed brand values do not coincide with the happenings on the ground, then more often than not, the resultant stakeholder actions prove to be considerably detrimental for the brand. "Employees have the potential to make or break the corporate brand" (Ind, 1998). Burmann and Zeplin (2005) proposed two behavioral constructs, namely, the "brand citizenship behavior" and the "brand commitment" in order to explain the manner in which employees help in maintaining a uniform brand identity. The former construct refers to all the forms of employee behavior, whether brand oriented or otherwise, which communicates the brand identity, while the latter refers to the psychological processes underlying the former. For instance, when an employee goes beyond the call of duty to endorse the brand in nonofficial situations (such as referring the brand to a friend or a relative) out of her commitment to the core brand values, she is displaying brand citizenship behavior originating out of her brand commitment. This behavior will occur only when the employee's core values and cultural orientations are congruent with the organization's core values and culture. If employees with core values and cultural personalities similar to the corporation's core brand values and cultural personality can be hired, then it would become much easier for organizations to promote and augment their corporate brands. This is where HRM department needs to collaborate with the marketing department and make sure that the personnel with the right set of core values and personalities are recruited.

Management needs to integrate marketing, organizational behavior and HRM functions of the organization in such a manner that the employees are better able to assimilate the brand and thereby become better brand ambassadors promoting the brand in a desirable manner to the customers or other stakeholders (Biswas & Suar, 2016; Burmann & Zeplin, 2005; Ind, 1998; Punjaisri& Wilson, 2017; Vallaster & Chernatony, 2005). Customers experience a brand through various points of contact and not only through the marketing personnel. For instance, a prospective car buyer interested in a particular brand due to its claim of being indigenous, may be put off if she happens to overhear employees of the car company talking about imported parts being used in manufacturing the car. This example may seem quite far-fetched, but there is no denying the fact that numerous such communication channels through employees are always open. For instance, if an employee, who shares the core brand values and believes in the vision of the brand, comes across some negative discussions about the company on a social media platform, she is likely to take initiative and defend the brand, while an employee who does not identify with the fundamental corporate brand values and thinks that the vision of the company is a sham, is likely to take no action or in a worst case scenario may add fuel to the fire. Burmann and Zeplin (2005) argued that the choices which employees make while dispensing their job responsibilities and their manner of functioning have a huge impact on the foundations of brand identity. Therefore, it becomes important for the managers to familiarize the employees with the core brand identity concepts so as to make sure that they appreciate the brand vision and the core values behind the brand. It is incumbent upon the marketing and the HRM departments to ensure that brand promise is synchronized with the internal communication, the marketing processes, and the branding strategies (Potgieter & Doubell, 2020). A company wishing to leverage the power of its employees in developing a successful corporate brand needs to imbue them with the core brand values.

SIGNIFICANCE OF HAVING COMMITTED EMPLOYEES FOR A CORPORATE BRAND

Organizations across the globe have started to recognize the importance of employees as creators, promoters, and protectors of corporate brands. The actions (or inactions) of the employees determine whether an organization will be able to achieve its corporate branding objectives or not. This makes it essential for corporations to formulate an employer branding policy enabling them to entice and employ the best workers available in the job market, while at the same time enabling them to utilize that strategy for strengthening their corporate brands. Aggerholm, Andersen, and Thomsen (2011) postulated that creation of efficacious employer brands necessitates fostering of novel sustainable relationships between the organization and its existing and prospective employees which are developed with the employees' support and cooperation. Modern corporations are dependent upon employees to propagate their vision, culture, and image amongst various stakeholders whether external or internal. "A strong brand identity that is lived by the employees is based on values that are congruent with those of the corporate culture" (Burmann & Zeplin, 2005, p. 293). The primary difference between corporate branding and product branding is that the former originates from inside of the organization and focuses more upon the internal stakeholders with the purpose of projecting the brand vision, values, culture, and image to the internal as well as the external stakeholders in a better and more efficacious manner, while the latter is more externally oriented. "Corporate branding necessitates a different management approach. It requires greater emphasis on factors internal to the organization, paying greater attention to the role of employees in the brand building process" (Harris & de Chernatony, 2001, p. 441). Employees are the most important instruments in creating a powerful corporate brand and therefore, it becomes imperative for the HRM department to focus on hiring those employees who share the brand values and vision of the organization. Also, the members of the brand management team should be able to coordinate well with the employees from the other departments so as to facilitate the communication of a unified brand to the external stakeholders (Harris & de Chernatony, 2001). A brand promise is propagated through multifarious communication media and experienced via the products and services offered by the corporation. However, none of these is more important than the behavior of the staff towards the external as well as the internal stakeholders. Balmer & Gray (2003) argued that it is incumbent upon the senior management to perspicaciously articulate the corporate branding proposition in such a manner that the brand promise is fulfilled for all the stakeholders.

Integrating HRM and corporate branding or what role HRM can play in the development of corporate brands is a relatively lesser explored area. Employer Branding has recently attracted the attention of the people in the academia as well as in the corporate world when it was felt that much of the branding effort had external orientation while the employees at large within the organization remained alienated with an organization's branding programs. As the major economies of the world shifted their focus from manufacturing to services, companies realized that employees play a substantial role in building a positive or a negative brand image in the eyes of the customers in particular and in societies in general. In a technology driven world where creativity and innovation are important drivers of success, it is the human capital which is strategically valuable and possessing unique talent can enable the organization to attain distinct competence over its competitors (Martin, 2009). Employer branding can enable the organization to attract those employees whose values are identical to those of the organization and help in developing brand loyalty among the existing employees through affective commitment. Schneider (1987) proposed that individuals tend to get attracted to those organizations who they perceive as having value congruence, that is, the value system of individuals is same or similar to the value system of the organization. Moreover, when organizational values are internalized by the employees, it creates an emotional connect between employee and the organization which further stimulates the employee performance. When an organization successfully creates its own identity, it can project itself as different from the rest and when the individual employees adopt these values, it helps in building a stronger employer brand, and consequently stronger corporate brand.

It is important to understand that in this era of continuous information bombardment and rapidly occurring technological changes, the employees or the potential candidates have access to multiple channels of communication. The companies are actively using various media to connect to prospective as well as existing employees. Internal branding's purpose is to sell the organization's value proposition to the internal stakeholders by communicating a unified and consistent brand image throughout an employee's work lifecycle (Sartain, 2005). Employer branding as a communication strategy can be used to project the organization as "workplace-of choice" to both the current and prospective employees. Employer branding works in consonance with internal branding strategies as well as external branding strategies. An exploratory research by Foster et al., (2010) suggested that the linkage between internal branding and employer branding leads to existing staff becoming brand ambassadors for the company. Internal branding creates better brand value congruence as employees internalize the organizational values and enhances employees' identification with the firm (Baker, Rapp, Meyer & Mullins, 2014) which leads to intellectual and emotional engagement of employees with the brand (Foster et al., 2010). These intellectually and emotionally engaged employees are invaluable for any organization as they are intrinsically motivated to promote the corporate brand. However, the role of HRM department as a strategic contributor to an organization still has not been given the importance it deserves (Martin, 2009; Srimannarayana, 2010). It is important for corporations to understand that HR needs to be taken on board in order to formulate policies to develop strong corporate brands.

The top companies actively seek to build a distinct organizational culture besides offering unique products and services to its customers. An organization with a strong culture may be able to elicit higher levels of motivation, loyalty, and performance from its employees. The right concoction of "differentiating capabilities" of employees and the organization's unique "cultural identity" helps it in attracting the people with the right person-job fit and in delivering unparalleled products and services due to the hired employees' high sense of pride and commitment to the organization (Mosley, 2014). Hatch and Schultz (2001) suggested that management of corporate brands requires integration of vision, culture, and image of an organization and this requires continuous communication amongst various stakeholders. According to them, organizational culture can be used as a catalyst to build effective corporate brands. Organizational culture plays a critical role in a firm's success as it provides an organizational identity to employees. It also gives them clarity about the vision of the organization and what it represents. Organizational culture shapes the brand image both for internal and external stakeholders (Obasan, 2012). Values and beliefs are central to the understanding of any organizational culture. Numerous studies have proved that integration of organizational and employee values contribute to facilitate congruent brand perceptions and results in more coherent brand identity (Harris & de Chernatony, 2001) and may lead to attainment of competitive advantage over other firms. Organizational culture can become a source of competitive advantage only if there is a fit between brand values and core values of employees and only then employees can contribute in the crafting of "corporate brand value" (Hatch & Schultz, 2001, p.1049). The challenge before the contemporary organizations is that the organizational boundaries have extended due to the continuous interaction between organizations and their multiple stakeholders which calls for a concurrence between the corporation's vision and the values espoused by their employees (Hatch & Schultz, 1997). In the global economy, a company hires people from different countries, which renders it necessary for human resource managers to understand that how these cultural differences influence employee the perception of the employer brand (Gehrels, 2019).

Employees may convey internal organizational issues to the customers and can have a strong impact on consumers' perception of the brand and the organization. A genuine corporate brand originates from those inherent cultural values of an organization which generate the symbolic connotations associated with the organization by various stakeholders (Hatch & Schultz, 2001). To build on these cultural values, companies need support of their employees. The HR department needs to become a strategic partner in the corporate brand building process. Projection of successful corporate brand amongst internal as well as external stakeholders would require an "inside-out" approach. This "inside-out" approach cannot be put into practice without the support of the employees. To maximize the participation of employees in the process of corporate branding, an integrated approach is required. This is where the role of employer branding attains importance. Garas, Mahran, and Mohamed (2018) carried out a study on 400 frontline bank employees in Egypt and found out that employees' brand supporting behavior may help the banks in building powerful corporate brands. They also suggested that "enhancing employees' role clarity and affective commitment will ensure sustainable brand supporting behaviour." Hoppe (2018) suggested that "symbolic job offerings may lead to "favorable brand-related employee attitudes and behaviours."

The employees should be made aware of the core brand values of the organization before they start to think about working with the company. This is imperative because if the core values of the organization are not in sync with the core values of the employee, it would become very difficult for them to participate in the corporate branding process. Employer branding has dual impact on an organization as it not only helps corporations to attract and retain the best talent, but also to cultivate a workforce that understands the core organizational values and culture of the organization. Cultivation of organizational culture and values as a component of corporate brand in the minds of the employees is not as simple a task as it may appear. One of the basic mistakes that companies make is that they focus too much on creation of external brand images while in the process creating something which is completely incongruent to what is going on within the organization. This leads to a gap in employees' perception of the organization and the projected image of the organization. Since, the employees do not believe in that image, they are unable to carry it forward to the external stakeholders, especially customers. The issue lies in the fact that organizations have not yet started to realize the full potential of employees as brand ambassadors or brand differentiators. The HR department needs to step-up as a strategic partner in building and managing corporate brands. It needs to formulate policies keeping in mind the interrelationships among various components of corporate branding such as corporate identity, corporate reputation, and corporate image (Martin, 2009). Formulations of such HR policies would help the organizations in realizing the true potential of their employees as value creators for their corporate brand.

Employer brand building process involves employees imbibing certain behavior like being courteous, responsive, reliable, helpful, and empathetic as these values are perceived important by customers while evaluating service quality and have an important bearing on consumer loyalty and retention (Miles & Mangold, 2004). The role of employees in creating and managing the brand image cannot be undermined as they form an important source of customer information. Organizations need to involve them as "ambassadors" of their brands in order to gain the employees' commitment for building the brand identity (Harris & de Chernatony, 2001). The top companies are applying the concept of employer branding on the lines of corporate branding (Mosley, 2015). Their focus has been on alluring and retaining the best employees who can be made to imbibe the core values of the organization in order to make them brand ambassadors of the corporate brand. The employees of the organization are the best source of positive communication about the brand to the customers. The employee who can identify herself with the cultural values of the organization can convey it best to other relevant stakeholders in the organization. The employee who internalizes employer brand can successfully build a successful brand–customer relationship (Xiong et al., 2013). Apart from communicating with external stakeholders, an employee may also act as brand ambassador for the internal employees, communicating the vision, and values of the organization to them. Their mission involves sharing (a) authentic and trustworthy information, (b) best practices followed by the organization, and (c) providing employee feedback and giving recommendation to the brand management team for improving the brand image (Schmidt & Baumgarth, 2018). The employees play a vital role in creating, developing, and promoting corporate branding and therefore, employer branding practices should be formulated and carried out in such a manner to enable the organizations to hire the employees whose cultural orientations and personal values are in consonance with the corporate brand.

CONCLUSION

This chapter has attempted to delineate the role of employees as an indispensable component of the corporate branding process. Corporate branding should adopt a bottom-to-top approach instead of opting for top-to-bottom because the latter creates a gap in the conception of corporate identities as perceived by the top management and all other stakeholders, especially employees. It is very important to involve employees at every step of the corporate branding process. Corporate brands are products of corporate vision, mission, culture, identity, image, and reputation. All these concepts get manifested in the functioning of the employees. If they are not aware of what their organization stands for or what are its values, they would not be able to carry it forward. Also, if their values are not aligned with the corporate's brand image, they would find it difficult to participate in the corporate brand building process. This is where employer branding can play a crucial role. Through careful employer branding, companies may attract the employees with not only the right skill set but with a perspicuous understanding of the company's core brand values. Since, such employees would already have favorable attitude toward the brand and the values it epitomizes; it would be easier for them to represent it. The companies, therefore, need to focus on employer branding not only as a tool to attract and retain the best talent, but also as a component of the corporate branding process. Happy employees mean happy customers, many organizations have begun to align their employer and consumer brand strategies (Mosley, 2015). Employer branding should be considered as a strategic tool and should be used to support the overall corporate branding process if the corporations wish to develop and promote strong corporate brands.

REFERENCES

Aaker, D. A. (1996). *Building Strong Brands*. New York: The Free Press.

Aaker, D. A. (2004). Leveraging the corporate brand. *California Management Review*, 46(3), 6–18. doi: https://doi.org/10.1177/000812560404600301

Abimbola, T., Lim, M., Foster, C., Punjaisri, K., & Cheng, R. (2010). Exploring the relationship between corporate, internal and employer branding. *Journal of Product & Brand Management*, 19(6), 401–409. doi: https://doi.org/10.1108/10610421011085712

Abimbola, T., Trueman, M., Iglesias, O., Abratt, R., & Kleyn, N. (2012). Corporate identity, corporate branding and corporate reputations. *European Journal of Marketing*, 46(7/8), 1048–1063. doi: https://doi.org/10.1108/03090561211230197

Aggerholm, H. K., Andersen, S. E., & Thomsen, C. (2011). Conceptualising employer branding in sustainable organizations. *Corporate Communications: An International Journal*, 16(2), 105–123. doi: https://doi.org/10.1108/13563281111141642

Ambler, T., & Barrow, S. (1996). The employer brand. *Journal of Brand Management*, 4(3), 185–206. doi: https://doi.org/10.1057/bm.1996.42

App, S., Merck, J., & Buttgen, M. (2012). Employer branding: Sustainable HRM as a competitive advantage in the market for high-quality employees. *Management Revue*, 23(3), 262–278. doi: www.jstor.org/stable/41783721

Backhaus, K. (2016). Employer branding revisited. *Organization Management Journal*, 13(4), 193–201. doi: https://doi.org/10.1080/15416518.2016.1245128

Backhaus, K., & Tikoo, S. (2004). Conceptualizing and researching employer branding. *The Career Development International*, 9(5), 501–517. doi: https://doi.org/10.1108/13620430410550754

Baker, T. L., Rapp, A., Meyer, T., & Mullins, R. (2014). The role of brand communications on front line service employee beliefs, behaviours, and performance. *Journal of the Academy of Marketing Science*, 42(6), 642–657. doi: https://doi.org/10.1007/s11747-014-0376-7

Balmer, J. M. (1995). Corporate branding and connoisseurship. *Journal of General Management*, 21(1), 24–46. doi: https://doi.org/10.1177/030630709502100102

Balmer, J. M. (2001). Corporate identity, corporate branding and corporate marketing-Seeing through the fog. *European Journal of Marketing*, 35(3/4), 248–291. doi: https://doi.org/10.1108/03090560110694763

Balmer, J. M., &Soenen, G. B. (1999). The acid test of corporate identity management™. *Journal of Marketing Management*, 15(1–3), 69–92. doi: https://doi.org/10.1362/026725799784870441

Balmer, J. M. T., & Gray, E. R. (2003). Corporate brands: what are they? What of them? *European Journal of Marketing*, 37(7/8), 972–997. doi: https://doi.org/10.1108/03090560310477627

Bhatnagar, J., & Srivastava, P. (2008). Strategy for staffing: Employer branding and person organization fit. *Indian Journal of Industrial Relations*, 44(1), 35–48. www.jstor.org/stable/27768170

Biraghi, S., & Gambetti, R. C. (2015). Corporate branding: Where are we? A systematic communication-based inquiry. *Journal of Marketing Communications*, 21(4), 260–283. doi: https://doi.org/10.1080/13527266.2013.768535

Biswas, M. (2013). Employer branding: A human resource strategy. In R. K. Pradhan & C. K. Poddar (Eds.), *Human Resources Management in India: Emerging Issues and Challenges* (pp.160–180). New Delhi: New Century Publications. Retrieved from www.researchgate.net/publication/307599173_Employer_Branding_A_Human_Resource_Strategy

Biswas, M. K., & Suar, D. (2016). Antecedents and consequences of employer branding. *Journal of Business Ethics*, 136(1), 57–72. doi: https://doi.org/10.1007/s10551-014-2502-3

Brønn, P. S. (2002). Corporate Communication and the Corporate Brand. In P. S. Brønn, & R. Wigg (Eds.), *Corporate Communication: A Strategic Approach to Building Reputation.* Oslo: GyldendalNorskForlag.

Burmann, C., & Zeplin, S. (2005). Building brand commitment: A behavioural approach to internal brand management. *Journal of Brand Management*, 12(4), 279–300. doi: https://doi.org/10.1057/palgrave.bm.2540223

Chang, A., Chiang, H., & Han, T. (2012). A multilevel investigation of relationships among brand-centered HRM, brand psychological ownership, brand citizenship behaviours, and customer satisfaction. *European Journal of Marketing*, 46(5), 626–662. doi: https://doi.org/10.1108/03090561211212458

Christensen, L. T., Morsing, M., & Cheney, G. (2008). *Corporate Communications: Convention, Complexity and Critique.* London: Sage Publications Ltd.

Collins, C. J. (2007). The Interactive effects of recruitment practices and product awareness on job seekers' employer knowledge and application behaviours. *Journal of Applied Psychology*, 92(1), 180–190. doi: https://doi.org/10.1037/0021-9010.92.1.180

De Chernatony, L., Harris, F., & Riley, F. D. O. (2000). Added value: Its nature, roles and sustainability. *European Journal of Marketing*, 34(1/2), 39–56. doi: https://doi.org/10.1108/03090560010306197

Dowling, G. (2001). *Creating Corporate Reputations: Identity Image, and Performance.* Oxford: Oxford University Press.

Edlinger, G. (2015). Employer brand management as boundary-work: A grounded theory analysis of employer brand managers' narrative accounts. *Human Resource Management Journal*, 25(4), 443–457. doi: https://doi.org/10.1111/1748–8583.12077

Edwards, M. R. (2010). An integrative review of employer branding and OB theory. *Personnel Review*, 39, 5–23. doi: https://doi.org/10.1108/00483481011012809

Foster, C. Khanyapuss, P., & Cheng, R. (2010). Exploring the relationship between corporate branding, internal branding and employer branding. *Journal of Product and Brand Management*, 19(6), 401–409. doi: https://doi.org/10.1108/10610421011085712

Garas, S. R. R., Mahran, A. F. A., & Mohamed, H. M. H. (2018). Internal corporate branding impact on employees' brand supporting behaviour. *Journal of Product & Brand Management*, 27(1), 79–95. doi: https://doi.org/10.1108/JPBM-03-2016-1112

Gehrels, S. (2019). *Employer Branding for the Hospitality and Tourism Industry: Finding and Keeping Talent*. Bingley, UK: Emerald Publishing Limited. doi: 10.1108/978-1-78973-069-220191001

Gelb, B. D., & Rangarajan, D. (2014). Employee contributions to brand equity. *California Management Review*, 56(2), 95–112. doi: https://doi.org/10.1525/cmr.2014.56.2.95

Hansen, J. H. (2012). *Is Social Integration Necessary for Corporate Branding. A Study of Corporate Branding Strategies at Novo Nordisk*, Doctoral Thesis, Copenhagen Business School, Frederiksber. PhD series, No. 22.2012. Retrieved from: http://libsearch.cbs.dk/primo_library/libweb/action/dlDisplay.do?docId=CBS01000567285&vid=CBS&afterPDS=true

Harris, F. & Chernatony, L. (2001). Corporate branding and corporate brand performance. *European Journal of Marketing*, 35(3/4), 441–456. https://doi.org/10.1108/03090560110382101

Hatch, M. J. & Schultz, M. (1997). Relations between organizational culture, identity and image. *European Journal of Marketing*, 31(5/6), 356–365. doi: https://doi.org/10.1108/eb060636

Hatch, M. J. & Schultz, M. (2001). Are the strategic stars aligned for your corporate brand. *Harvard Business Review*, 79(2), 128–134. Retrieved from: https://hbr.org/2001/02/are-the-strategic-stars-aligned-for-your-corporate-brand

Hatch, M. J. & Schultz, M. (2003). Bringing the corporation into corporate branding. *European Journal of Marketing*, 37(7/8), 1041–1064. doi: https://doi.org/10.1108/03090560310477654

Hoppe, D. (2018). Linking employer branding and internal branding: Establishing perceived employer brand image as an antecedent of favourable employee brand attitudes and behaviours. *Journal of Product & Brand Management*, 27(4), 452–467. doi: https://doi.org/10.1108/JPBM-12-2016-1374

Ind, N. (1998). An integrated approach to corporate branding. *Journal of Brand Management*, 5(5), 323–329. doi: https://doi.org/10.1057/bm.1998.20

Interbrand Best Global Brands 2019. (2019). Retrieved from www.interbrand.com/best-brands/best-global-brands/2019/ranking/ (accessed on 04/08/2020).

Kane, C. (2015). By popular demand: Companies that changed their ways. *CNBC*, 28 April 2015, www.cnbc.com/2015/04/27/by-popular-demand-companies-that-changed-their-ways.html.

Kapfere, J. N. (2002). Corporate brand and organizational identity. In B. Moingeon & G. Soenen (Eds.), *Corporate and Organizational Identities: Integrating Strategy, Marketing, Communication and Organizational Perspectives* (pp. 175–194). London: Routledge.

Kennedy, S. H. (1977). Nurturing corporate images. *European Journal of marketing*, 11(3), 119–164. doi: https://doi.org/10.1108/EUM0000000005007

King, S. (1991). Brand-building in the 1990s. *Journal of Marketing Management*, 7(1), 3–13. doi: https://doi.org/10.1080/0267257X.1991.9964136

Knox, S., & Bickerton, D. (2003). The six conventions of corporate branding. *European Journal of Marketing*, 37(7/8), 998–1016. doi: https://doi.org/10.1108/03090560310477636

Kreps, G. (1981). *Organizational Folklore: The Packaging of Company History*. RCA Paper Presented at the ICA/SCA Conference on Interpretive Approaches to Organizational Communication. Alta. UT.

Kucherov, D. & Zamulin, A. (2016). Employer branding practices for young talents in IT companies (Russian Experience). *Human Resource Development International*, 19(2), 178–188. doi: https://doi.org/10.1080/13678868.2016.1144425

Laforet, S. (2017). Effects of organizational culture on brand portfolio performance. *Journal of Marketing Communications*, 23(1), 92–110. doi: https://doi.org/10.1080/13527266.2014.956230

Louis, M. R. (1980). *A Cultural Perspective on Organizations: The Need for and Consequences of Viewing Organizations as Culture-Bearing Milieux*. Paper Presented at the National Academy of Management Meetings, Detroit, MI.

Martin, G. (2009). Driving corporate reputations from the inside: A strategic role and strategic dilemmas for HR? *Asia Pacific Journal of Human Resources*, 47(2), 219–235. doi: https://doi.org/10.1177/1038411109105443

Martin, G., Gollan, P. J., & Grigg, K. (2011). Is there a bigger and better future for employer branding? Facing Up to innovation, corporate reputations and wicked problems in SHRM. *The International Journal of Human Resource Management*, 22(17), 3618–3637. doi: https://doi.org/10.1080/09585192.2011.560880

Melewar, T. C., Gotsi, M., & Andriopoulos, C. (2012). Shaping the research agenda for corporate branding: avenues for future research. *European Journal of Marketing*, 46(5), 600–608. doi: https://doi.org/10.1108/03090561211235138

Melewar, T. C., Gotsi, M., Andriopoulos, C., Fetscherin, M., & Usunier, J. C. (2012). Corporate branding: An interdisciplinary literature review. *European Journal of Marketing*, 46(5), 733–753. doi: https://doi.org/10.1108/03090561211212494

Meyer, A. (1981). How ideologies supplement formal structures and shape responses to environments? *Journal of Management Studies*, 19, 45–61. doi: https://doi.org/10.1111/j.1467-6486.1982.tb00059.x

Miles, S. J., & Mangold, G. (2004). A conceptualization of the employee branding process. *Journal of Relationship Marketing*, 3(2–3), 65–87. doi: https://doi.org/10.1300/J366v03n02_05

Mitchell, A. (1997), *Brand Strategies in the Information Age, Financial Times Report*. London: Financial Times Retail & Consumer.

Morsing, M. (2006). Corporate moral branding: Limits to aligning employees. *Corporate Communications: An International Journal*, 11(2), 97–108. doi: https://doi.org/10.1108/13563280610661642

Mosley, R. (2014). Employer brand management: *Practical Lessons from the World's Leading Employers*. West Sussex, UK: John Wiley & Sons Ltd.

Mosley, R. (2015). CEOs need to pay attention to employer branding. *Harvard Business Review*. 11. Retrieved from: https://hbr.org/2015/05/ceos-need-to-pay-attention-to-employer-branding

Obasan, K. A. (2012). Organizational culture and its corporate image: A model juxtaposition. *Business and Management Research*, 1(1), 121–130. doi:10.5430/bmr.v1n1p121

Olins, W. (1978). The corporate personality: An inquiry into the nature of corporate identity. London: Design Council, p. 212.

Olins, W. (2000). Why Brands are taking over the Corporation. In S. Majken, M. J. Hatch, & M. H. Larsen (Eds.), *The Expressive Organization: Linking Identity, Reputation, and the Corporate Brand* (pp. 51–65). Oxford: Oxford University Press.

Palazzo, G., & Basu, K. (2007). The ethical backlash of corporate branding. *Journal of Business Ethics*, 73, 333–346. doi: https://doi.org/10.1007/s10551-006-9210-6

Potgieter, A., & Doubell, M. (2020). The influence of employer branding and employees' personal branding on corporate branding and corporate reputation. *African Journal of Business and Economic Research*, 15(2), 109–135. doi: 10.31920/1750-4562/2020/v15n2a6

Punjaisri, K., & Wilson, A. (2017). The role of internal branding in the delivery of employee brand promise. In Balmer J. M. T., Powell S. M., Kernstock J., Brexendorf T. O. (Eds.),

Advances in Corporate Branding. Journal of Brand Management: Advanced Collections. London: Palgrave Macmillan. doi: https://doi.org/10.1057/978-1-352-00008-5_6

Saini, G., Rai, P. & Chaudhary, M. (2014). What do best employer surveys reveal about employer branding and intention to apply? *Journal of Brand Management*, 21, 95–111. doi: https://doi.org/10.1057/bm.2013.10

Sartain, L. (2005). Branding from the inside out at Yahoo: HR'S role as brand builder. *Human Resource Management*, 44(1), 89–93. doi: https://doi.org/10.1002/hrm.20045

Schall, M. S. (1981). *An Exploration into a Successful Corporation's Saga-Vision and its Rhetorical Community.* Paper Presented at the ICA/ SCA Conference on Interpretive Approaches to Organizational Communication, Alta, UT.

Schmidt, H. J., & Baumgarth, C. (2018). Strengthening internal brand equity with brand ambassador programs: Development and testing of a success factor model. *Journal of Brand Management*, 25(3), 250–265. doi: https://doi.org/10.1057/s41262-018-0101-9

Schneider, B. (1987). The people make the place. *Personnel Psychology*, 40, 437–453. doi: https://doi.org/10.1111/j.1744–6570.1987.tb00609.x

Schultz, M., & de Chernatony, L. (2002). The challenges of corporate branding. *Corporate Reputation Review*, 5(2/3), 105–114.

Siehl, C. & Martin, J. (1981). Learning organizational culture, Working Paper, Graduate School of Business, Stanford University.

Srimannarayana, M. (2010). Human resource roles in India, *Indian Journal of Industrial Relations*, 46(1), 88–99. www.jstor.org/stable/25741099

Tanwar, K. & Kumar, A. (2019). Employer brand, person-organization fit and employer of choice: Investigating the moderating effect of social media. *Personnel Review*, 48(3), 799–823. doi: https://doi.org/10.1108/PR-10-2017-0299

Turban, D. B. & Greening, D. W. (1997). Corporate social performance and organizational attractiveness to prospective employees. *Academy of Management Journal*, 40, 658–672. doi: https://doi.org/10.5465/257057

Vallaster, C. & de Chernatony, L. (2005). Internationalisation of services brands: The role of leadership during the internal brand building process. *Journal of Marketing Management*, 21(1–2), 181–203. doi: https://doi.org/10.1362/0267257053166839

Vallaster, C. Lindgreen, A. & Maon, F. (2012). Strategically leveraging corporate social responsibility: A corporate branding perspective. *California Management Review*, 54(3), 34–60. doi: https://doi.org/10.1525/cmr.2012.54.3.34

Wilson, A. M. (2001). Understanding organizational culture and the implications for corporate marketing. *European Journal of Marketing*, 35(3/4), 353–367.

Xiong, L., Kinga, C., & Piehler, R. (2013). "That's not my job": Exploring the employee perspective in the development of brand ambassadors. *International Journal of Hospitality Management*, 35, 348–359.

Yakimova, R., Mavondo, F., Freeman, S., & Stuart, H. (2017). Brand champion behaviour: Its role in corporate branding. *Journal of Brand Management*, 24(6), 575–591.

4 Toward an Integration of the Collaborator's Experience in the Digital Management of the Employer Brand

Zakaria Lissaneddine, Mostapha El Idrissi, and Younès El Manzani

INTRODUCTION

The management of the employer brand is increasingly attracting interest from companies that rely on their human capital to generate competitive advantage (App, Merk, & Büttgen, 2012). The attraction and retention of human resources can be achieved through a captivating employer brand (Rana et al., 2019, Sharma et al., 2019). In this perspective, Mosley (2014) underlines that the management of the employer brand (sourcing, recruitment, integration and talent management, etc.) has a significant impact on the competitiveness of the company more than any other activity of the HR function (Brymer, Molloy, & Gilbert, 2014; Carpentier et al., 2017; Ha & Luan, 2018).

At the same time, social media are occupying an important place in the business world more than ever. The rapid evolution of social media has significantly transformed managerial practices (Otken & Okan, 2016). The HR function is considered to be one of the organizational dimensions that has been able to take advantage of these digital platforms to strengthen organizational attractiveness. Indeed, social media are no longer only perceived as a means of entertainment, but also as real competitive tools; thereby organizations, aware of this reality, exploit social media in a perspective of management of the employer brand (Ha & Luan, 2018; Otken & Okan, 2016; Sharma & Verma, 2018). Compared to traditional media, social media are more effective, as it allows not only to communicate with employees but also to interact with potential candidates to promote awareness of the company as an employer (Kissel & Büttgen, 2015; McFarland & Ployhart, 2015).

The literature that is interested in studying the relationship between employer brand and social medias (Carpentier et al., 2017; Carpentier, Van Hoye, & Weng, 2019; Eger, Mičík, Gangur, & Řehoř, 2019; Kashive, Khanna, & Bharthi, 2020; Kissel & Büttgen, 2015; Mičík & Mičudová, 2018; Otken & Okan, 2016; Sivertzen, Nilsen, & Olafsen, 2013; Tanwar & Kumar, 2019) remains fragmented and there is a lack of research that studies the entire employer brand management process as well as the role that collaborators may play in it. This chapter, therefore, proposes to explore the management of the employer brand via social media in the customer relationship management sector in Morocco, where the employee's turnover rate can vary between 15% and 25% depending on the nature of the company's activity.[1] More specifically, the study reported by this chapter mobilized an exploratory interview guide with managers of four call centers who use social media to attract potential candidates while involving their employees in digital ambassadorship strategies.

TOWARD DIGITALIZATION OF EMPLOYER BRAND MANAGEMENT

The advent of the Internet has fundamentally changed recruiting practices and made it possible for recruiters to reach new candidates. As individuals have adapted to web trends, businesses have also started to use social media for a variety of purposes, including promoting their employer brand. Social media are considered to be one of the most popular means of communication to implement HR marketing strategies (Bondarouk, Ruël, Axinia, & Arama, 2014; Carpentier et al., 2017; Eger et al., 2019; El Zoghbi & Aoun, 2016; Mičík & Mičudová, 2018; Otken & Okan, 2016).

Today, many social media offer opportunities for businesses to create, share, and develop content in a participatory and collaborative way (Kaplan & Haenlein, 2010). To differentiate themselves from product marketing websites, HR managers are increasingly deploying digital strategies where the employer brand is honored. After developing his strategy and positioning, the HR Marketing manager has a range of Web Marketing levers, including social media, to apply his action plan and highlight the attributes of the employer brand. To do this, social media are used to share information with job applicants about the company's internal culture, career prospects, and benefits. Companies tend to develop positive and desirable attractiveness based on official information (Turban & Cable, 2003), but also on unintentionally provided information that comes from various sources such as social media (Bondarouk et al., 2014; Girard, Fallery, & Rodhain, 2011; Kissel & Büttgen, 2015).

Social media offer new uses to company managers in terms of information sharing, digital monitoring, processing of stakeholder requests, and identification of new opportunities. The use of these digital technologies has become very widespread in the field of HR marketing (Kluemper, Mitra, & Wang, 2016). Previously, organizations used social media to market their brand image to external stakeholders such as customers, suppliers, and shareholders. Nowadays, social media offer the company the possibility of creating a positive employer brand and good organizational attractiveness when it shares precise and complete information with potential candidates on its internal culture, work climate, career opportunities, and development prospects (Bondarouk et al., 2014; El Zoghbi & Aoun, 2016; Turban & Cable, 2003). According to Carpentier et al. (2017), companies that use social media to convey relevant information sought by job seekers have an employer brand image that appears more compelling and appealing, which drives the interest and the intent to apply for potential candidates who wish to join these companies.

The challenge of managing the employer brand via social media is to serve the interests of the organization in terms of internal and external HR marketing, to attract the best candidates who may exist in the work market. On the other hand, this strategy aims to unveil the sense of belonging among employees, encourage them to share and communicate the values of the organization in public to promote the employer brand of the company. Social media also allows employers to post job vacancies by leaning on their employees, who can share these ads online with their networks and promote their business. This is a digital version of word of mouth, but faster, on a larger scale, and more geographically dispersed (McFarland & Ployhart, 2015).

Employer brand management can, therefore, take advantage of new digital platforms to attract potential candidates, interact with employees, and obtain valuable information on a targeted category of Internet users. Thus, while offering employer's very useful solutions in terms of sourcing, recruitment, and e-reputation.

A STUDY IN CUSTOMER RELATIONSHIP CENTERS IN MOROCCO

Context and Field of Research

The employer brand remains an emerging concept in Morocco; however, its popularity among Moroccan managers is constantly growing. More specifically, HR marketing practices are spreading rapidly in the offshore customer relationship management segment. Considered, rightly or wrongly, as the *new factories of the future* (Buscatto, 2002), call centers to arouse the interest of researchers and professionals (Makkaoui, 2012). Much of the literature on customer relationship centers focuses on management styles, employee's well-being, and working conditions. While the issues of management of the employer brand within these structures tend to be neglected.

New trends in employee's attraction and retention in the relationship management industry such as sharing employee's experience and interacting with potential candidates on social media as well as employer brand ambassadorship aroused our interest, because of their particularity which consists in mobilizing social media throughout the management process of the employer brand.

TABLE 4.1
List of the Interviewed Managers

Interviewee	Function	Seniority	Duration	Firms acronyms	Location
LK	HR Marketing & Social Media Manager	15 years	2H 05	WH	Rabat
YA	HR Project Manager	11 years	1H 10		
SM	HR Marketing Officer	10 months	1H 35		
IK	Sourcing & Employer Brand Officer	2 years	55 Min		
SB	Recruitment Officer	6 years	1H 05	IN	Casablanca
KH	Social Media Manager	6 months	55 Min		
HB	Internal Communications Officer	2 years	1H 45		
ME	Communication Manager	1,5 years	1H 15	AS	
SS	HR Development Manager	4 years	1H 25		
JS	Communications Officer	7 years	55 Min		
MB	Deputy Director of Communication	3 years	1H	PG	
AM	Digital Communication Project Manager	10 months	45 Min		

To better understand these new managerial practices relating to the management of the employer brand, qualitative data was collected through 12 semi-structured interviews conducted with managers of four call centers. These companies alone employ 40% of the employees belonging to the customer relationship management segment in Morocco. While knowing that this same segment of activity employs more than 75,000 people[2] and includes more than 90 call centers[3] located in 14 different cities of the country. The interviews took place during the period from August until October 2017 to obtain the opinion of the 12 managers. The function and affiliation of the interviewees are specified in more detail in Table 4.1.

From a Digital Employer Branding Mastered by Companies ...

The employer brand occupies a very important place in the managerial language of the companies observed in this study, the people questioned attribute to the employer brand a strategic aspect which is equivalent to the overall notoriety of the company given the nature of the sector of activity which is considered very sensitive. Within these companies, the tasks of managing the employer brand are mainly entrusted to the Human Resources department, but this does not prevent close collaboration with other departments such as the communication and marketing departments. The Human Resources department of call centers is generally divided into several working groups, focused on different axes, such as recruitment, HR marketing, HR projects, sourcing, and employer branding.

> Digital marketing and the employer brand, all of this comes under the HR department, but there is also the communication (department), we work in close collaboration with the communication department, the only difference between us is that we are more focused on the outside: e-recruitment, digital communication and social networks, and the communication department is more aimed at internal employees and external customers of the company.
>
> (Sourcing and Employer Brand Officer – WH)

Companies are increasingly vigilant of all sources of information that potential candidates go-to for information about their future employer, even before they apply (Benraïss–Noailles, Lhajji, Benraïss, & Benraïss, 2016). Potential candidates use social media to search for information about the quality of life within the company and the position to be filled.

> For me, the employer brand is the image that the company can reflect, it is also word of mouth effect because if someone has already had an experience here within AS, then he will certainly tell it to those around him and his family and so that's where the influence of the employer brand begins in my opinion.
>
> (HR Development Manager – AS)

The proactivity of the companies that are the subject of our research has led them to manage their employer brand digitally. They use social media to communicate on all the benefits enjoyed by their current employees, thus making potential candidates want to join them (Viot, Benraïss–Noailles, Herrbach, & Benraïss, 2015). The leaders of the companies observed underlined the importance of social media in their communication strategy, particularly in the promotion of their employer brand.

> With the establishment of the Digital Employer Brand (DEB) unit in 2016, the group's management broadened my scope of intervention concerning everything related to the employer brand, whether it be e-recruitment websites, social media management or e-reputation monitoring ....
>
> (HR Marketing & Social Media Manager – IN)

The process of the employer branding in call centers begins with the value proposition and the improvement of the business image. The value proposition represents a package that can include fair compensation, a good work climate, social and economic benefits, to meet the needs of the employee and guarantee him a good employee experience within the company.

> An attractive employer brand is a pleasant working climate, well-appointed premises, understanding managers, first of all, it is the human aspect that makes the company and not the people, you can have the best of all companies, but if your manager is not good or if you are not comfortable in your work you will not like the company ….
>
> (Social Media Manager – IN)

Based on the interviews carried out, the managers claimed that their companies were a pleasant place to work within which a good atmosphere prevails, favorable working conditions, and many opportunities for professional development. Those conclusions go hand in hand with our theoretical framework. When these companies wish to demonstrate that they are a great place to work in and thus promote their employer brand, they communicate their values and their culture and try to disseminate the daily experiences of their employees during work to create a coherent image between their employer brand and the reality within the company (Backhaus & Tikoo, 2004; Edwards, 2009).

The second step of the employer brand management process is employee loyalty. In call centers, the retention of employees has become a top management priority, more and more of these structures realize that human capital is now one of the most valuable intangible assets they own (Goujon–Belghit, Gilson, & Bourgain, 2015). Employee's commitment is an important goal for employer brand management (Rana & Sharma, 2019), the factors that encourage employee's commitment should be considered when developing the employer's value proposition (Mosley, 2014). Giving a sense to the tasks performed by employees can make their jobs attractive. Thus, fulfillment at work, training, recognition, and motivation of employees would lead to organizational involvement.

> A good employer brand is a brand that succeeds in being attractive to its employees in the first place before being attractive to its customers, for me a brand always starts from the inside, regardless of its field of activity, the most important thing is to serve its employees. I often say that what is experienced internally is seen externally …
>
> (Communication Manager – AS)

Retention of employees within companies is based on the nonexistence of gaps in employee's perception before and after recruitment. In other words, due to the information collected mainly on digital social networks, potential candidates conceive a certain image of their future employer before joining the company. The adequacy of this perception designed by future employees with the reality experienced within the company facilitates the integration of the later and their retention.

> We are already on four social networks: Facebook, LinkedIn, Instagram, and YouTube. In the last two, I put them aside because they are mainly media. So, our approach is that we have two messages between Facebook and LinkedIn: on Facebook, we have a fun, offbeat, playful message … We position ourselves in a younger target and we focus more on the customer advisor profile, so we are going to talk a lot about our internal life in the company, events that can

> be challenging, meetings, breakfasts, birthdays, celebrations… on LinkedIn we adopt a different editorial line, we publish all our support business offers and we have a more corporate message on LinkedIn than on Facebook, that's how we deploy our employer brand on social networks, we reflect our internal life, we are not in the artificial discourse, we reflect what is happening in our sites.
>
> (Digital Communication Project Manager – PG)

The interviewed managers deploy their company's digital strategy in terms of HR marketing through a digital presence on various networks and social media. The operationalization of this strategy is carried out according to a specific communication style to each social network: mass communication on Facebook, which targets young job seekers who have the profile of call center agents, and corporate and segmented communication on LinkedIn which targets middle management profiles. The content of this communication mainly deals with aspects related to employee experience and working conditions within the company.

DSNs are mobilized to attract and communicate with potential candidates throughout the recruitment process, they are used to share the daily experiences of employees with potential candidates through direct and spontaneous dissemination (diffusion) in images and videos, these actions enable future employees to discover the work climate within the company, by highlighting the social and financial advantages of the latter. This allows employer brand managers to communicate and broadcast messages online to a specific group of potential candidates who best meet the company's recruiting needs. In this sense, proximity and active listening to potential candidates make the company an attractive employer, thanks to social media, call centers can promote their reputation as an employer by pointing up their strengths such as management style, social commitment and meeting the needs of employees, to be recognizable and become responsible social actors, corporate citizens and above all employers of choice.

> When we say that we are the leader of customer relationship centers in Morocco, we must prove that we are number one in Morocco, through our daily actions, our social responsibility, our diversity charter, our HR policy, and our management … Afterward, we can accept dissatisfied people, and this is what makes the strength of our employer brand, it is listening and be listened to by people who are dissatisfied and who criticize. Those people represent only a minority, which confirms that our employer brand has been attractive and has been attentive to suggestions for improvement.
>
> (HR Marketing Officer – WH)

… to the Integration of Employees into the Digital Employer Branding

According to Ababneh, LeFevre, and Bentley (2019), organizations that invest in employee engagement will reap significant benefits in employee's productivity, the achievement of organizational goals, customer satisfaction, and talent retention. Organizational loyalty is achieved through meeting employee's needs and delivering on employer brand promises.

The managers interviewed during this study claim to have participated in the promotion of the employer brand of their company in different ways. The recommendation of the company with its entourage and participation in social and philanthropic activities initiated by the company are the main forms by which employees promote the employer brand of their companies.

> When I set up an e-recruitment website, the first population with whom I must communicate before I share it on Facebook are my collaborators, I will send a nice email to all the staff saying "good news, today I am announcing the creation of our new WH career e-recruitment website, your role is to share this information as much as possible with those around you, our watchword is like and share." My internal communication is based on employee promoters so that they are my ambassadors externally since they are more much credible than a marketing representative or a communication manager who will share this via Facebook … The ambassadors are the pillars of the success of the employer brand, a company that does not have internal ambassadors, its employer brand is not good and healthy, it has more shortcomings than strengths.
>
> (HR Marketing and Social Media Manager – WH)

The use of social media could also promote the effect of inbound marketing (Megargel, Shankararaman, & Reddy, 2018) for the dissemination and promotion of the employer brand. This strategy consists of attracting Internet users to the company's website and official pages and supporting them in a communication tunnel to transform the Internet user into a potential candidate, the potential candidate into an employee and the employee into an ambassador.

In this sense, Schlager, Bodderas, Cachelin, and Maas (2011) advocate that the employer brand would have a positive impact on employees, who once satisfied, could influence the experience of both customer and potential candidates. The initiatives of ambassador employee can generate a feeling of attachment to the organization among other employees (Sahu, Pathardikar, & Kumar, 2018) and help disseminate a positive image of the employer to potential candidates (Schlager et al., 2011). This deduction can be applied to the case of employees in the customer relationship sector in Morocco, the latter can improve the attractiveness of their sector and positively develop its business image. Customer relationship centers that practice this strategy can be the exception and become employers of choice by enhancing the value of the teleoperator profession.

The war for talent (Beechler & Woodward, 2009; Charbonnier–Voirin & Vignolles, 2016; Kolesnicov, 2017; Michaels, Handfield-Jones, & Axelrod, 2001; Somaya & Williamson, 2008) is a visible phenomenon in several sectors that encourages many organizations to take care of their brand employer to be more attractive and reduce their turnover rate by retaining the most competent employees (Charbonnier-Voirin, Laget, & Vignolles, 2014). Thus, having an attractive and credible employer brand makes it easier to target and attract the right profiles for the customer relationship management profession, which limits "casting errors" among call centers. Maintaining a positive image of the company as an employer of choice can be considered as a solution to fight turnover in customer relations centers in Morocco.

> You have to know how to retain the people you have, as they are the ones who can propel you forward, if each time you train employees, you include them in production, they gain experience, but after that cannot retain them, they will leave and repeating this process will generate a cost and the results will not be good, so it is better to keep the resources that we have as staff, to retain them, after when we will have other recruits, it will be to move towards other objectives and avoid turnover and not to replace those who have left.
>
> (Recruitment Officer – IN)

The digital management of the employer brand would then be a process of continuous improvement, it does not stop after the achievement of a given objective, but it is a capitalization of achievements, it consists in taking care of the brand employer to make the company's image more attractive on digital platforms. An employer brand that meets the expectations and ambitions of employees in terms of social benefits and working conditions would minimize departures and establish a sense of belonging among employees. Once satisfied, they would become ambassadors of their employer brand, which constitutes proof of credibility for potential candidates wishing to join the company. According to Kapoor (2010), employer brand management can cope with labor shortages.

Social media serves as an indicator to assess a company's digital reputation and gauge the expectations of potential candidates. Thus, paying attention and maintaining the employer brand comes under e-reputation monitoring (Fueyo, 2015). Contacting Internet users on digital social networks as part of community management allows HR managers to measure the basis of criticisms and objections to deal with them on the web or notify the department concerned via reporting to remedy problems raised during these online exchanges. The e-reputation intelligence on DSNs is also used to perform benchmarks with competitors to discover new good practices and specially to avoid bad ones by taking advantage of the experience of rivals in the field, all this with the aim of continuous improvement of the employer brand. The aforementioned tasks are carried out through reporting containing information collected from the company's official DSN pages. This report is then analyzed to learn lessons and useful implications for the care and promotion of the company's reputation as an employer.

> We closely follow everything that is said about us, it allows us to know how we are being followed, what young job seekers are asking for … We have monitoring tools to see what is happening, to follow up on all events in case there is a reputation crisis or a bad buzz. We are there, we are alerted, we do not let things blow up in front of us. We work also in partnership with a public relations agency which allows us to receive daily reports concerning all the sources that cite our box on the internet. We also receive the "best practices" of our competitors.
>
> (Social Media Manager – IN)

DISCUSSION

Considered as an emerging concept, employer brand is nowadays the main facet of HR marketing and refers to all the advantages communicated by an organization to make it an attractive and referenced workplace (Charbonnier–Voirin, Marret, & Paulo, 2017). In this context, firms have invested different tools to promote their reputation as employers among potential candidates. In addition to traditionally using tools such as the company's website, job boards or business forums, a growing number of organizations are now using social media platforms such as LinkedIn, Viadeo, Facebook, or Twitter in their policy of attraction and recruitment (Girard et al., 2011) and are considering the creation or rather the formalization of an ambassadors' network (Charbonnier-Voirin & Vignolles, 2015). Thus, it seems that a strong employer brand associated with a well thought digital strategy is a promising opportunity to acquire and engage the talents within organizations (Bondarouk et al., 2014; Girard et al., 2011; Kissel & Büttgen, 2015). Lessard (2015, p. 103) indicates that "Firms can no longer ignore the scope and importance of social media in their communication strategy, notably in disseminating their employer brand." Through this research, we attempted to gain a better insight into the employer brand management practices in the digital era. Therefore, our ambition was to contribute to the comprehension of the process of the employer brand management in the Moroccan organizational context and more particularly in the customer relationship management segment.

Employer brand remains an essential tool for the development of the human capital of every company, and more particularly for those operating in the customer relationship management segment that we've studied in this research. This activity segment is characterized by high volatility in the workforce, generally attributed to the characteristics of the millennial generation (Soulez & Guillot–Soulez, 2011) and also to the nature of the activity itself. Achieving a credible employer brand makes it easier to acquire the right talent for call center positions, and social networks could have a crucial role in disseminating it to potential candidates and job seekers.

From the side, continuous improvement of the employer brand as well as the respect of the promises the firm made to its employees lead to retaining the latter (Chaminade, 2010; Kapoor, 2010). The commitment and the loyalty of the employees then translate in the expression of pride of belonging. Thus, employees tend to identify with their company notably by sharing their daily work life with their circle in social networks (Helm, 2011; van Zoonen, Bartels, van Prooijen, & Schouten, 2018). Sharing and disseminating the "employee experience" turns into a spontaneous way of promoting the employer brand of a firm. Ambassadorship can also take the form of sponsoring recruits which makes it easier to hire, integrate, and retain these recruits (Kapoor, 2010).

Communicating on the various aspects related to call center positions in the context of employer brand management could be an effective remedy for excessive employee turnover. The functional benefits of the employer brand include the working conditions and climate, in addition to the nature and characteristics of the professional activity in question. Credible and objective communication about the aforesaid aspects will attract the right profiles for customer relationship management professions. In the same vein, promoting the employer brand of call centers through various channels and particularly via social networks will help enhance the business image of these structures.

Potential advantages of a digital strategy of employer brand management have been mentioned by several authors (Bondarouk et al., 2014; Girard et al., 2011; Kissel & Büttgen, 2015; Wolf, Sims, & Yang, 2015). Reducing recruitment costs is the key advantage of using these digital technologies by the Human Resources Department because then the recruitment process becomes shorter and more flexible. Indeed, employers get into direct contact with job seekers without having to go through an intermediate (recruitment agencies) or sometimes expensive advertising (communication agencies). In addition to joining and interacting with a large community of internet users, social networks put a spotlight on employees which eventually allows a humanization of the company. Integrating social networks in the range of HR sourcing tools will also optimize the process of dissemination and development of the employer brand. Figure 4.1 puts the initial results of our exploratory research on call centers into perspective.

The advantages offered by social networks in terms of employer brand management are accompanied by challenges related to a lack of control over these

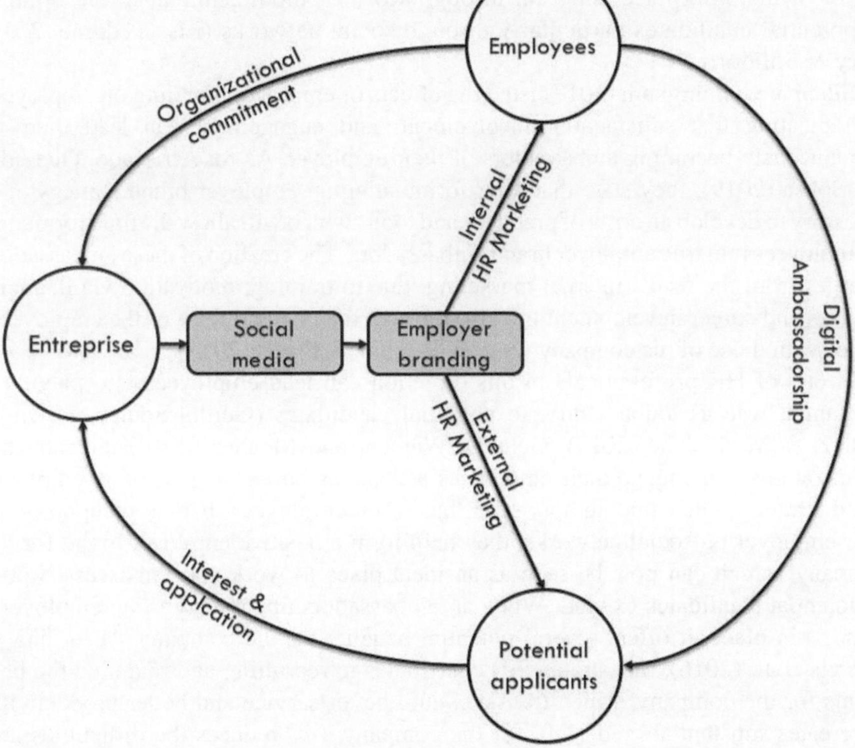

FIGURE 4.1 Proposal of a conceptual framework of the employer brand's digital management. (Sources: Authors' own work.)

new communication channels, which sometimes leads to misinformation and bad buzz. Therefore, a significant time budget must be dedicated to digital monitoring. This requires a certain level of dedication and availability of managers all the time so as not to allow space for e-reputation crises. This context incites companies to be permanently present on several digital platforms to manage virtual communities, continuously brush up their identity and reputation online. Digital monitoring, which consists of observing and evaluating interactions on social media, requires careful monitoring and instant reactivity. To do this, an initial financial investment is necessary for the company to be equipped with the essential technologies to perform the aforementioned tasks.

Brand ambassadorship is still a very recent phenomenon in the business world (Andersson & Ekman, 2009; Quaratino & Mazzei, 2018; van Zoonen et al., 2018). The expressions "living the brand," "brand champions," and "brand ambassadors" are terms used in the literature to refer to high performing employees who bring the brand to life through their attachment and devotion to it (Wallace & de Chernatony, 2009). Today's employees play a greater role in executing employer brand processes (Lissaneddine, Chaouki, & Rodhain, 2018). Thus, an employer brand ambassador is perceived as a credible witness to the distinctive and attractive nature of his workplace and can, through word of mouth, influence the opinion on potential candidates particularly through social networks (Lissaneddine, 2019; Parry & Solidoro, 2013).

Gilani & Cunningham (2017)listed the effects of employer branding on employees by indicating that satisfaction, involvement, and engagement can lead them to spontaneously becoming ambassadors of their employer. As for Arriscado, Quesado, and Sousa (2019), they state that after formulating an employer brand strategy, it is necessary to develop an array of practices and tools, which will allow the transformation of employees into true employer brand ambassadors. The creation of these ambassadors emerges from the firm's internal marketing, due to training, motivation via financial bonuses and career development but also through the compatibility of the employee's values with those of his company (Näppä, Farshid, & Foster, 2014).

Efforts of HR professionals in this direction can lead employees into playing a substantial role as ambassadors to potential candidates (Guillot-Soulez, St-Onge, Soulez, & Merkouche, 2018). Gehrels, Wienen, and Mendes (2016) indicate that organizations can engage their employees as "ambassadors" as part of an employer brand strategy. The same authors state that when employees display recognition of their employer on social networks, they help form a positive employer brand for the company, which can now be seen as an ideal place to work and thus sees its pool of potential candidates expand. When an ambassadorship strategy of an employer's brand is in place, it offers several potential benefits for the company. According to Gehrels et al. (2016), this strategy (i) contributes to recruiting and retaining the best talents for the company, which translates into better service and better productivity, (ii) creates substantial credibility for the company, (iii) reduces the difficulties and costs of the recruitment process, (iv) increases the number of suitable candidates, and (v) improves the company's reputation.

With the advent of digitalization, each employee shares content: he can express his opinion, comment, interact about his company, all without prior approval from the latter (Chaintreuil, 2015). Thus, each collaborator participates in shaping the image that the company conveys in terms of its employer brand. Interactions with current and potential employees are therefore becoming faster, more frequent, and above all more transparent. In this way, the consistency between the employer brand offer and the reality experienced in the company can be ensured, which leads to strengthening the credible dissemination network made up of "ambassadors" who create direct and less institutional communication. Chaintreuil (2015) indicates that in the digital era, issues related to the employer brand are evolving. Indeed, social networks today represent a tremendous opportunity to build an employer brand image. The HR function has a major role in this change because its mission is to involve employees in a global strategy of employer brand ambassadorship and to ensure that they become real ambassadors for the company.

Electronic word of mouth and digital traces generated by employees on the social web are crucial in communication policies (Fueyo, 2015). More precisely, the use of the digital voice of employees is crucial in the brand ambassadorship strategy. Graham and Cascio (2018) studied the impact of employer branding on the company's organizational reputation while focusing on the role of employees as employer brand ambassadors of their own company. The same authors reported that social media are used as a channel for spreading positive word-of-mouth about their employer. Thus, ambassadors tend to express their satisfaction, loyalty, and pride of belonging to social networks.

CONCLUSION

The main purpose of this study was to provide a better understanding of what employer brand management via social networks can bring to companies in terms of organizational attractiveness and loyalty. As in any research work, this study had several limitations, particularly related to the research field, which is small and consists only of four companies. A sample as small as this does not sufficiently describe the digital management practices of the employer brand, suggesting that the results cannot be generalized to other settings.

In light of the limits and the various contributions of this research work, several areas of study or extension can be considered. This research represents a starting point for future studies, as it is one of the first to ever address the issue of the employer brand management via social networks. It is recommended for future research to refine the theoretical framework that feeds the current research subject and thus enriches the literature with new theoretical and conceptual proposals. A first perspective would be to carry out a study with a much larger sample composed of companies from different activity sectors, which are characterized by a digital presence on social media. This would allow us to globally understand the extent of the use and mobilization of social networks for employer brand management.

Certainly, the people who were questioned in our study represent big companies struggling with problems of attraction and retention due to the business image of their activity sector. However, the purpose of this research deserves to be developed and deepened with different stakeholders in the employer brand management process, namely potential candidates. It would be particularly interesting to understand how these candidates perceive organizations that seek to appeal and attract them, and also the determining factors and aspects that make an employer brand attractive to job seekers. Another interesting study track would be to focus more on the company's employees since they are the intermediates between managers and potential candidates. The aim would be to conduct a very thorough survey of this audience's perspective.

The results of this research raise the need for further studies in other activity sectors that are experiencing the same trends in managing and attracting human resources. Future research in other industries such as banking, for example, may confirm the results and conclusions of our research.

NOTES

1 Study carried out by the National Telecommunications Regulatory Agency, available for consultation from the follow ing link: www.anrt.ma/sites/default/files/Etudes_dev_centres_appel_maroc_2007_0.pdf

2 Newspaper article "lesecos.ma" of October 5, 2017, available at: http://leseco.ma/business/60463-offshoring-la-croissance-est-au-rendez-vous.html

3 Website specialized in the trades of call centers in Morocco, available at: www.moncallcenter.ma/centres-appels.php

REFERENCES

Ababneh, O. M. A., LeFevre, M., & Bentley, T. (2019). Employee engagement: development of a new measure. *International Journal of Human Resources Development and Management*, *19*(2), 105. https://doi.org/10.1504/IJHRDM.2019.098623

Andersson, M., & Ekman, P. (2009). Ambassador networks and place branding. *Journal of Place Management and Development*, *2*(1), 41–51. https://doi.org/10.1108/17538330910942799

App, S., Merk, J., & Büttgen, M. (2012). Employer branding: Sustainable HRM as a competitive advantage in the market for high-quality employees. *Management Revue*, *20*(1), 53–69. https://doi.org/10.2307/41783721

Arriscado, P., Quesado, H., & Sousa, B. (2019). Employer branding in the digital era attracting and retaining millennials using digital media. In M. Túñez-López, V. Martínez-Fernández, X. López-García, X. Rúas-Araújo, & F. Campos-Freire (Eds.), *Communication: Innovation & Quality. Studies in Systems, Decision and Control,* 154, 391–403. https://doi.org/10.1007/978-3-319-91860-0_23

Backhaus, K., & Tikoo, S. (2004). Conceptualizing and researching employer branding. *Career Development International*, *9*(5), 501–517. https://doi.org/10.1108/13620430410550754

Beechler, S., & Woodward, I. C. (2009). The global "war for talent." *Journal of International Management*, *15*(3), 273–285. https://doi.org/10.1016/j.intman.2009.01.002

Benraïss-Noailles, L., Lhajji, D., Benraïss, A., & Benraïss, B. (2016). Impact de la réputation classique et de l'e-réputation sur l'attractivité des entreprises en tant qu'employeurs. *Question(s) de Management*, *15*(4), 71–80. https://doi.org/10.3917/qdm.164.0071

Bondarouk, T., Ruël, H., Axinia, E., & Arama, R. (2014). What is the future of employer branding through social media? Results of the Delphi study into the perceptions of HR professionals and academics. In Tanya Bondarouk and Miguel R. Olivas-Luján (eds.). *Social Media in Human Resources Management* (pp. 23–57). Bingley: Emerald Publishing Limited. https://doi.org/10.1108/S1877-6361(2013)0000012006

Brymer, R. A., Molloy, J. C., & Gilbert, B. A. (2014). Human capital pipelines. *Journal of Management*, *40*(2), 483–508. https://doi.org/10.1177/0149206313516797

Buscatto, M. (2002). Les centres d'appels, usines modernes? Les rationalisations paradoxales de la relation téléphonique. *Sociologie Du Travail*, *44*(1), 99–117. https://doi.org/10.1016/S0038-0296(01)01202-X

Carpentier, M., Van Hoye, G., Stockman, S., Schollaert, E., Van Theemsche, B., & Jacobs, G. (2017). Recruiting nurses through social media: Effects on employer brand and attractiveness. *Journal of Advanced Nursing*, *73*(11), 2696–2708. https://doi.org/10.1111/jan.13336

Carpentier, M., Van Hoye, G., & Weng, Q. (2019). Social media recruitment: Communication characteristics and sought gratifications. *Frontiers in Psychology*. https://doi.org/10.3389/fpsyg.2019.01669

Chaintreuil, J. N. (2015). *RH & Digital: Regards collectifs de RH sur la transformation digitale*. Paris: Diateino.

Chaminade, B. (2010). *Attirer et fidéliser les bonnes compétences: Créer votre marque employeur* (AFNOR). La Plaine Saint-Denis.

Charbonnier-Voirin, A., Laget, C., & Vignolles, A. (2014). L'influence des écarts de perception de la marque employeur avant et après le recrutement sur l'implication affective des salariés et leur intention de quitter l'organisation. *Revue de Gestion Des Ressources Humaines*, *93*(3), 3–17. https://doi.org/10.3917/grhu.093.0003

Charbonnier-Voirin, A., Marret, L., & Paulo, C. (2017). Les perceptions de la marque employeur au cours du processus de candidature. *Management & Avenir*, *94*(4), 33. https://doi.org/10.3917/mav.094.0033

Charbonnier-Voirin, A., & Vignolles, A. (2015). Marque employeur interne et externe: Un état de l'art et un agenda de recherche. *Revue Française de Gestion*, *41*(246), 63–82. https://doi.org/10.3166/rfg.246.63-82

Charbonnier-Voirin, A., & Vignolles, A. (2016). Enjeux et outils de gestion de la marque employeur: point de vue d'experts. *Recherches En Sciences de Gestion*, *112*(1), 153. https://doi.org/10.3917/resg.112.0153

Edwards, M. R. (2009). An integrative review of employer branding and OB theory. *Personnel Review*, *39*(1), 5–23. https://doi.org/10.1108/00483481011012809

Eger, L., Mičík, M., Gangur, M., & Řehoř, P. (2019). Employer branding: Exploring attractiveness dimensions in a multicultural context. *Technological and Economic Development of Economy*, *25*(3), 519–541. https://doi.org/10.3846/tede.2019.9387

El Zoghbi, E., & Aoun, K. (2016). Employer branding and social media strategies. In F. Ricciardi & A. Harfouche (Eds.), *Information and Communication Technologies in Organizations and Society Past, Present and Future Issues* (pp. 277–283). Cham, Switzerland: Springer International Publishing. https://doi.org/10.1007/978-3-319-28907-6_18

Fueyo, C. (2015). *E-réputation corporate: influence de la voix digitale des employés via l'e-contenu de leur « Profil » sur les réseaux sociaux professionnels; application au secteur automobile.* Université Toulouse 1 Capitole.

Gehrels, S., Wienen, N., & Mendes, J. (2016). Comparing hotels' employer brand effectiveness through social media and websites. *Research in Hospitality Management*, *6*(2), 163–170. https://doi.org/10.1080/22243534.2016.1253282

Gilani, H., & Cunningham, L. (2017). Employer branding and its influence on employee retention: A literature review. *The Marketing Review*, *17*(2), 239–256. https://doi.org/10.1362/146934717X14909733966209

Girard, A., Fallery, B., & Rodhain, F. (2011). L'apparition des medias sociaux dans l'e-GRH: gestion de la marque employeur et e-recrutement. *16ème Congrès de l'AIM*. Saint Denis, Ile de la Réunion. Retrieved from https://hal.archives-ouvertes.fr/hal-00843689

Goujon-Belghit, A., Gilson, A., & Bourgain, M. (2015). Repenser les liens entre la gestion du capital humain et la marque employeur perçue en contexte de mutation organisationnelle. *Gestion et Management Public*, *3*(1), 53–71. https://doi.org/10.3917/gmp.033.0053

Graham, B. Z., & Cascio, W. F. (2018). The employer-branding journey: Its relationship with cross-cultural branding, brand reputation, and brand repair. *Management Research: Journal of the Iberoamerican Academy of Management*, *16*(4), 363–379. https://doi.org/10.1108/MRJIAM-09-2017-0779

Guillot-Soulez, C., St-Onge, S., Soulez, S., & Merkouche, W. (2018). L'identité coopérative comme élément distinctif de la marque employeur: une étude comparative France/Québec dans le secteur financier. *29ème Congrès de l'AGRH*. Lyon, France. Retrieved from www.researchgate.net/publication/329308972_L'identite_cooperative_comme_element_distinctif_de_la_marque_employeur_une_etude_comparative_FranceQuebec_dans_le_secteur_financier

Ha, N. M., & Luan, N. V. (2018). The effect of employers' attraction and social media on job application attention of senior students at pharmaceutical universities in Vietnam. *International Journal of Business and Society*, *19*(2), 473–491. Retrieved from www.ijbs.unimas.my/index.php/content-abstract/current-issue/485-the-effect-of-employers-attraction-and-social-media-on-job-application-attention-of-senior-students-at-pharmaceutical-universities-in-vietnam

Helm, S. (2011). Employees' awareness of their impact on corporate reputation. *Journal of Business Research*, *64*(7), 657–663. https://doi.org/10.1016/j.jbusres.2010.09.001

Kaplan, A. M., & Haenlein, M. (2010). Users of the world, unite! The challenges and opportunities of Social Media. *Business Horizons*, *53*(1), 59–68. https://doi.org/10.1016/j.bushor.2009.09.003

Kapoor, V. (2010). Employer branding: A study of its relevance in India. *IUP Journal of Brand Management*, *7*(1/2), 51–75.

Kashive, N., Khanna, V. T., & Bharthi, M. N. (2020). Employer branding through crowdsourcing: Understanding the sentiments of employees. *Journal of Indian Business Research*. https://doi.org/10.1108/JIBR-09-2019-0276

Kissel, P., & Büttgen, M. (2015). Using social media to communicate employer brand identity: The impact on corporate image and employer attractiveness. *Journal of Brand Management*, *22*(9), 755–777. https://doi.org/10.1057/bm.2015.42

Kluemper, D. H., Mitra, A. and Wang, S. (2016), "Social Media use in HRM", Research in Personnel and Human Resources Management (vol. 34). Emerald Group Publishing Limited, pp. 153–207. https://doi.org/10.1108/S0742-730120160000034011

Kolesnicov, I. (2017). *Winning the War for Talent: A Study on Employer Branding from a Corporate Communication Perspective*. Aarhus University.

Lessard, S. (2015). Communiquer la marque employeur sur les médias sociaux. *Gestion*, *40*(1), 100–103. https://doi.org/10.3917/riges.401.0100

Lissaneddine, Z. (2019). *Management de la marque employeur via les réseaux sociaux numériques: Cas des centres de la relation client au Maroc*. Cadi Ayyad University.

Lissaneddine, Z., Chaouki, F., & Rodhain, F. (2018). L'usage des réseaux sociaux numériques dans le management de la marque employeur: Cas des centres de la relation client au Maroc. *29ème Congrès de l'AGRH*. Lyon: France. Retrieved from www.researchgate.net/publication/326045328_L%27usage_Des_Reseaux_Sociaux_Numeriques_Dans_Le_Management_De_La_Marque_Employeur_Cas_Des_Centres_De_La_Relation_Client_Au_Maroc

Makkaoui, M. (2012). *GRH et Organisation du travail dans les centres d'appels délocalisés au Maroc*. Université de Liège.

McFarland, L. A., & Ployhart, R. E. (2015). Social media: A contextual framework to guide research and practice. *Journal of Applied Psychology*, *100*(6), 1653–1677. https://doi.org/10.1037/a0039244

Megargel, A., Shankararaman, V., & Reddy, S. K. (2018). Real-time inbound marketing: A use case for digital banking. In David Lee Kuo Chuen and Robert Deng (eds.) *Handbook of Blockchain, Digital Finance, and Inclusion, Volume 1* (pp. 311–328). Elsevier. https://doi.org/10.1016/B978-0-12-810441-5.00013-0

Michaels, E., Handfield-Jones, H., & Axelrod, B. (2001). *The war for talent* (H. B. S. Press, Ed.). Boston, MA: McKinsey & Company, Inc.

Mičík, M., & Mičudová, K. (2018). Employer brand building: using social media and career websites to attract Generation Y. *Economics & Sociology*, *11*(3), 171–189. https://doi.org/10.14254/2071-789X.2018/11–3/11

Mosley, R. (2014). *Employer brand management: Practical lessons from the world's leading employers*. Cornwall: Wiley.

Näppä, A., Farshid, M., & Foster, T. (2014). Employer branding: Attracting and retaining talent in financial services. *Journal of Financial Services Marketing*, *19*(2), 132–145. https://doi.org/10.1057/fsm.2014.9

Otken, A. B., & Okan, E. Y. (2016). The role of social media in employer branding. In M. Bilgin & D. H (Eds.), *Entrepreneurship, Business and Economics* (Vol. 13/1, pp. 245–260). https://doi.org/10.1007/978-3-319-27570-3_20

Parry, E., & Solidoro, A. (2013). Social media as a mechanism for Engagement? In Tanya Bondarouk and Miguel R. Olivas-Luján (eds.) *Social Media in Human Resources Management (Advanced Series in Management, Vol. 12)* (pp. 121–141). Bingley: Emerald Publishing Limited. https://doi.org/10.1108/S1877-6361(2013)0000012010

Quaratino, L., & Mazzei, A. (2018). Managerial strategies to promote employee brand consistent behavior. *EuroMed Journal of Business*, *13*(2), 185–200. https://doi.org/10.1108/EMJB-02-2017-0008

Rana, G., & Sharma, R. (2019). Assessing impact of employer branding on job engagement: A study of banking sector. *Emerging Economy Studies*, *5*(1), 7–21. https://doi.org/10.1177/2394901519825543

Sahu, S., Pathardikar, A., & Kumar, A. (2018). Transformational leadership and turnover. *Leadership & Organization Development Journal*, *39*(1), 82–99. https://doi.org/10.1108/LODJ-12-2014-0243

Schlager, T., Bodderas, M., Cachelin, J. L., & Maas, P. (2011). The influence of the employer brand on employee attitudes relevant for service branding: An empirical investigation. *Journal of Services Marketing*, *25*(7), 497–508. https://doi.org/10.1108/08876041111173624

Sharma, S., & Verma, H. V. (2018). Social media marketing: Evolution and change. In Githa Heggde and G. Shainesh (eds.) *Social Media Marketing* (pp. 19–36). Cham: Springer. https://doi.org/10.1007/978-981-10-5323-8_2

Sivertzen, A.-M., Nilsen, E. R., & Olafsen, A. H. (2013). Employer branding: Employer attractiveness and the use of social media. *Journal of Product & Brand Management*, *22*(7), 473–483. https://doi.org/10.1108/JPBM-09-2013-0393

Somaya, D., & Williamson, I. O. (2008). Rethinking the "War for Talent." *MIT Sloan Management Review*, *49*(4), 29–34. Retrieved from https://sloanreview.mit.edu/article/rethinking-the-war-for-talent/

Soulez, S., & Guillot-Soulez, C. (2011). Marketing de recrutement et segmentation generationnelle: regard critique a partir d'un sous-segment de la generation Y. *Recherche et Applications En Marketing*, *26*(1), 39–57. https://doi.org/10.1177/076737011102600103

Tanwar, K., & Kumar, A. (2019). Employer brand, person-organisation fit and employer of choice: Investigating the moderating effect of social media. *Personnel Review*, *48*(3), 799–823. https://doi.org/10.1108/PR-10-2017-0299

Turban, D. B., & Cable, D. M. (2003). Firm reputation and applicant pool characteristics. *Journal of Organizational Behavior*, *24*(6), 733–751. https://doi.org/10.1002/job.215

van Zoonen, W., Bartels, J., van Prooijen, A.-M., & Schouten, A. P. (2018). Explaining online ambassadorship behaviors on Facebook and LinkedIn. *Computers in Human Behavior*, *87*, 354–362. https://doi.org/10.1016/j.chb.2018.05.031

Viot, C., Benraïss-Noailles, L., Herrbach, O., & Benraïss, B. (2015). Attractivité organisationnelle et capital marque employeur: Une analyse par sous-dimensions. *33ème Université d'été de l'Audit Social et Gestion Des Ressources Humaines*, 1–16. Montréal. Canada. Retrieved from http://archives.marketing-trends-congress.com/2016/pages/PDF/VIOT_BENRAISS-NOAILLES_HERRBACH_BENRAISS.pdf

Wallace, E., & de Chernatony, L. (2009). Exploring brand sabotage in retail banking. *Journal of Product & Brand Management*, *18*(3), 198–211. https://doi.org/10.1108/10610420910957825

Wolf, M., Sims, J., & Yang, H. (2015). Look who's co-creating: Employer branding on social media. *ECIS 2015 Proceedings*, Paper 205. https://doi.org/10.18151/7217533

5 Employer Branding and Social Media

The Case of World's Best Employers

Megha Bharti and Anjuman Antil

INTRODUCTION

Today social networking service is the most efficient (costless & fast) and most effective way of creating relationships with various stakeholders for an organization. Social media allows users to interact by sharing and forming communities of mutual interest. There is a paradigm shift in human resource (HR) practices with the onset of social media as a tool for creating a digital word of mouth (Mičík & Mičudová, 2018). The HR managers believe that social networking platforms can give an additional advantage in attracting the right pool of candidates in coherence with the theory of person–organization fit (Bhatnagar & Srivastava, 2008; Tanwar & Kumar, 2019), effective employee engagement and collaborative communication. It can act as an instrumental tool of employer branding and signaling the brand image of the company (App et al., 2012) to gain a competitive advantage.

Despite the increasing use of social media in HR practices and relevance for the companies, there is not much existing literature that combines employer branding with social media on such a large scale (Tanwar & Kumar, 2019). This chapter considered six social networking platforms which are used by top employers of the world. The chapter also presented valuable implications for the employers in the context of employer branding strategies. Employer branding is recognized as a wide-ranging recruitment approach that places an organization in an appealing way in the minds of the potential candidates. A major portion of the "potential employees" category will be formed by the millennial in the near future, who are well equipped and habitual to the use of internet (Schawbel et al., 2015). This generation is characterized by different expectations than the previous generations (Hauw et al., 2010), implying that the companies cannot continue to use the old school strategies anymore. This warrants an in-depth investigation about the SNSs in creating competitive employer branding.

Social networking platforms are web-based dynamic platforms which help in information sharing, interactions and collaborations (Kashive et al., 2020). According to Kaplan and Hanenlein (2010) social media is a congregation of web applications based on the technology and ideological foundations of "Web 2.0," which allows development and exchange of user-generated content. The social media platforms can be categorized into four main groups (Grzesiuk & Wawer, 2018;Kaur, 2013): social networking sites like Facebook (networking site to connect with friends, family and communities), LinkedIn (employment oriented web service), Instagram (social networking service of sharing photos and videos);content sharing sites like YouTube (web video-sharing platform) which allows content to be uploaded in audio-visual mode with a scope for comments or discussion; Microblogs like Twitter which allows only a limited character writing on a verified profile which can also be accompanied by a picture or a video and collaborative projects like Wikipedia (edited and created by volunteers across the world) (Gehrels et al., 2016).

The objective of this chapter is to evaluate the presence of selected top employers on internet and social media, and to make recommendations for human resource managers on the effective use of social media as a strategic employer branding tool. For this, a study of top 50 companies on the Forbes best employers list 2019 is made. The analysis covers the presence of these organizations on internet, the manner in which they promote themselves online and the level of interactivity their presence involves, across selected social media platforms (Clair, 2016; Gehrels et al., 2016). This chapter investigates the factors that the employers focus on their online strategies for creating an attractive "Employer Branding Value Proposition" (Barrow & Mosley, 2011) by undertaking a qualitative and quantitative analysis using six social networking platforms – Corporate website, Facebook, Instagram, Twitter, YouTube and LinkedIn. They were scrutinized based on eight employer branding propositions (application value, social value, interest value, work–life balance, economic value, management value, development value and brand image (Reis & Braga, 2016; Dabirian et al., 2019).

Further analysis was done based on five important pre-selected thematic criterions form the literature, relevant in the context of effective employer branding, namely, (i) description of organization's culture and values (ii) account of learning and career development opportunities, (iii) Job vacancy search function (ease of navigation and transparency in updates of job vacancy and interviews) (iv) Posts pertaining to the employee achievements in the last month and (v) Posts pertaining to employee wellbeing initiatives. Sharing information about the company's organization culture as well as employee stories or achievements is a way to project an organization's identity, which can be an effective signaling tool for potential applications from self-congruity (Sirgy, 1982) and person–organization fit (Chatman, 1989; App et al., 2012) perspectives.

THEORETICAL BACKGROUND

The term "employer brand" was used for the first time in 1996 by Ambler and Barrow (Dabirian et al., 2019). According to them, employer brand can be defined as "the package of functional, economic, and psychological benefits provided by employment, and identified with the employing company." According to Berthon et al. (2005), employer branding is the aggregation of organization's efforts to convey existing and prospective employees that it is a desirable work place. Lievens et al. (2007) consider employer branding as a tool of corporate identity management by creating an image which is attractive and distinct. Overall, employer branding uses branding principles to attract new employees as well as engage, motivate and retain existing employees (Backaus & Tikoo, 2004; Collins & Kanar, 2013; Reis & Braga, 2016; Sharma R., Singh S. P., and Rana G. 2019). Employee branding has been measured by using Employee Attractiveness Scale of Berthon et al. (2005). The dimensions of this scale are derived from employer brand study of Ambler and Barrow's (1996). This scale has been widely used in the past by several researchers to measure the employer brand effectiveness (Roy, 2008; Arachchige & Robertson, 2011). This scale encompasses five dimensions: application value, development value, economic value, interest value and social value.

This research extends it further to include a total of eight dimensions (Reis & Braga, 2016) based on a resource-based view (Barney, 1991), self-congruity (Sirgy, 1982) and person–organization fit (Chatman, 1989; App et al., 2012) perspectives. In context of resource-based view, firm value is related to human capital and if managed skillfully it can lead to sustained competitive advantage (Barney, 1991; Dabirian et al., 2019). Though the resource-based view is very prevalent in the employer branding studies but the combination of these three perspectives is very rarely sought together (Sirgy, 1982; Bhatnagar & Srivastava, 2008). Bhatnagar and Srivastava (2008) and Tanwar and Kumar (2019) linked the person–organization fit with employer branding. There are no major studies to our knowledge mentioning the self-congruity effects concerning the relation between a job seeker's self-image and the employer's brand personality in the social media framework.

The following are the criterions considered for eight employer brand value propositions (see Table 5.1). A value proposition is the core message communicated by the company about its offering to the employees.

LITERATURE REVIEW

This section reviewed the literature relevant in the context of proposed research that is the employer branding value propositions and social networking platforms.

TABLE 5.1
Employer Brand Value Propositions

Employer Brand Value Propositions	Criterions
Application value	Humanitarian workplace
	Gives back to society
	Gives importance to diversity
	Customer orientation
	Opportunities for learning and applying knowledge
Interest value	Stimulating and Challenging work environment
	Encouragement and appreciation for creativity
	Innovative practices in work, product and services
Social value	Gratification felt by employees by working with others
	Cordial relation with colleagues and leaders
	Encouragement and support by colleagues and leaders
	Team approach
	Cheerful work environment
Economic Value	Good growth opportunities
	Attractive compensation
	Job Security
Development value	Increase in employee's confidence
	Work recognition and self-worth
	Career enhancement
Management value	Leadership capabilities of management
	Manager competence
	Strong Vision
	Ability to inspire and motivate
Work/Life balance	Volunteering
	Flexible work hours
	Leisure activities
	Family or other non-work-related activities
Brand Image	Perceived notion of target audience
	Overall persona

Source: Authors' extraction from the literature.

Researchers have been increasingly giving importance to the concept of social media-based employer branding (Reis & Braga, 2016; Dabirian et al., 2019). Table 5.2 gives a brief snapshot of the researches conducted in this area.

Dabirian et al. (2019) compared the reviews of the past and the current employees on Glassdoor and concluded the importance of employer brand value propositions. Tanwar and Kumar (2019) conducted a quantitative study using employer attractiveness scale focused on the effect of person–organization effect on employer of choice and the moderation effect of social media on the relation between these two. Saini (2020) took case of Shopper Stop, a retail store and studied its employee branding strategies using different social media platforms.

TABLE 5.2
Literature Review

Author (Year)	Sector/Sample	Platform	Methodology	Findings
Sivertzen et al. (2013)	Oil and gas industry Defense and maritime	Social Media Use	Quantitative analysis	Employer attractiveness values increase corporate reputation and recruitment rate
Kissel and Büttgen (2015)	30 largest German companies	Facebook	Quantitative analysis	Social media is critical for communicating brand meaning and attracting new talents
Gehrels et al. (2016)	Hospitality	LinkedIn Facebook Career Websites	Content Analysis	All the reviewed brand showed the possibility of improving online presence
Carpentier et al. (2017)	Belgian Hospitals	LinkedIn Facebook	Experiment	Nurse exposure to LinkedIn and Facebook had positive effect on employer brand values
Mičík and Mičudová (2018)	Top 60 Czech employers	LinkedIn Facebook Twitter YouTube Career Websites	Content Analysis	Facebook and Twitter are most popular platforms
Grzesiuk and Wawer (2018)	100 Largest Polish Private Companies	LinkedIn Facebook Twitter YouTube	Case Study Method	Companies were not using systematic approach of employer branding strategies
Dabirian et al. (2019)	IT	Glassdoor	Content Analysis	Learning, development opportunities and non-monetary perks are important for employees
Tanwar and Kumar (2019)	Top 10 employers	Social Media Use	Quantitative analysis	Social media affect the relationship among person–organization fit and employer of choice
Saini (2020)	Retail	LinkedIn Facebook Twitter YouTube	Content Analysis	Use of Social media to manage and promote employer brand
Kashive et al. (2020)	Pharma IT Retail FMCG	Glassdoor	Text and Sentiment Analysis	Themes generated were in line with employer value proposition. Social value is the most common value.

Source: Authors' extraction from the literature.

Sivertzen et al. (2013) analyzed the effect of employer brand values using social media on corporate reputation and recruitment rate and concluded a positive effect of social media usage on employer attractiveness. According to Kissel and Büttgen (2015), the usage of social media is very critical when an employer wishes to portray a particular brand identity in order to retain and attract talented people. Gehrels et al. (2016), Mičík and Mičudová (2018), and Dabirian et al. (2019) used content analysis to explore the employer branding strategies used by companies on different social networking platforms. They concluded that social media platforms help in building a strong corporate brand, publicize their customer and social activities, highlight new innovation products, increase global outreach and attract suitable talent.

Past researches have reviewed the concept of social media and employer branding considering either one sector with single platform (Dabirian et al., 2019; Robertson et al., 2019) or multiple platforms (Gehrels et al., 2016; Carpentier et al., 2017; Saini, 2020) or top employers of a specific country mostly on multiple platforms (Kissel & Büttgen, 2015; Mičík & Mičudová, 2018). In this chapter the researchers have considered top employers all around the world of different sectors along with five social media platforms and their corporate career website. This gives a vast coverage and in-depth knowledge of the important value or communication for employer branding.

DATA AND METHODOLOGY

In this chapter, the online employer brand presence and effectiveness of the top 50 companies on "Global 2000: The World's Best Employers" list (Forbes 2019) are compared. Data were collected across five social networking platforms (Facebook, Instagram, Twitter, YouTube, and LinkedIn) as well as the corporate website. Figure 5.1 shows the industrial based classification of the selected companies for the analysis:

Information was collected on 21 dimensions (Appendix 5.1) with regard to the six SNSs page statistics including the number of followers across the five social media platforms, update frequency (average total posts on these platforms in the last one month), number of employer branding posts out of the total posts of the last one month as well as the average likes on the latest five posts. Information on existence of a YouTube channel as well as the various elements of the company specific corporate website/career website, namely, (i) existence of a career and jobs section, (ii) listing of job vacancies, and (iii) inclusion of employee reviews/stories, were collected in a dichotomous fashion (yes/no).

Content analysis of company controlled social media platforms is a relevant research instrument because of the potential of the rapidly increasing online content and information in influencing a potential applicant's view on a company (Lai & To, 2015). A content analysis enables a comprehensive understanding of social presence of a company, while focusing on an integrated analysis of texts and their thematic content (Zhang & Wildemuth, 2009). With the help of quantitative and qualitative content analysis (Weber, 1990), the presence of a companies' employer brand and the value proposition of their employer brand message was retrieved in the form of

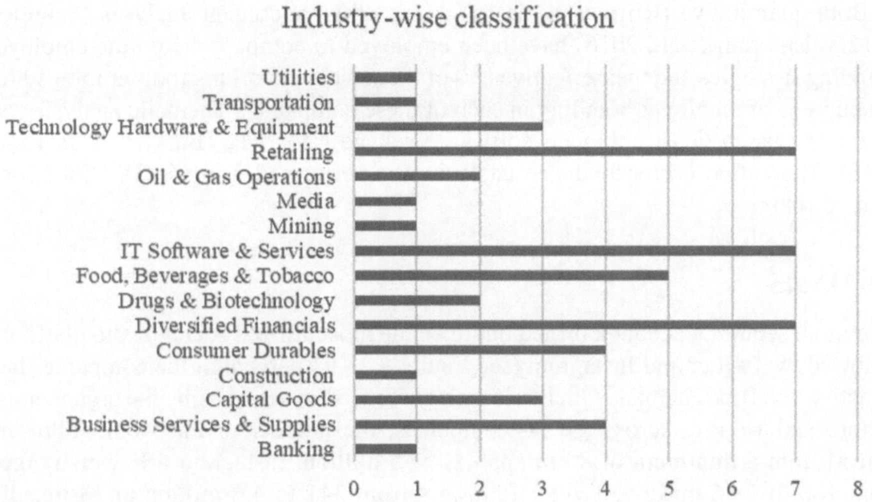

FIGURE 5.1 Industry-wise classification of companies.
(Source: Authors' analysis based on Forbes list.)

Facebook posts, Instagram captions, LinkedIn posts as well as Tweets written in the last one month, which were then assessed on the eight dimensions of the "Employer Branding Value Proposition" – application value, development value, interest value, social value, economic value work–life balance, management value and brand image (Dabirian et al., 2019).

The six SNSs specific to each company's brand were also scrutinized on five important criterions, relevant in the context of effective employer branding, namely, (i) description of organization's culture and values (ii) account of learning and career development opportunities, (iii) Job vacancy search function (ease of navigation and transparency in updates of job vacancy and interviews), (iv) Posts pertaining to the employee achievements in the last month, and (v) Posts pertaining to employee well-being initiatives.

A cumulative social media presence (SMP) score as well as an online employer branding (OEB) score was given to each of these 50 companies based on standardized coding of the retrieved quantitative data using SPSS V25. The coding was done on a scale of 1 to 5 (1 being the lowest score). Each of the 21 elements was scored out of 5 based on specific range criterions. The dichotomous variables were given a score of 1 for yes and 0 for no. Companies that had no posts in the last one month on a social networking site or had no employer branding posts out of the total one-month posts, were given a score of 0 on these dimensions. Three Chinese companies – GD Power Development (rank 21), WH Group (rank 25) and Founder Securities (rank 48) had no social media presence on any of the considered 5 platforms, thus were excluded from the analysis. This can be attributed to the ban of various global social media providers like Facebook and Google in mainland China.

Both quantitative (Riffe et al., 2019) and qualitative content analyses (Schreier, 2012;Vaismoradi et al., 2016) have been employed to compare the online employer branding strategies and their effectiveness of the social networking tools employed for the purpose of employer branding in each of these companies. Thematic analysis was performed using QDA computer software package NVivo 12 (Bazeley & Jackson, 2013). Thereafter, correspondence analysis was performed using SPSS V25 (Doey et al., 2011).

ANALYSIS

Across all sectors, Facebook turned out to be the most utilized social media platform, followed by Twitter and Instagram (see Figure 5.2). 91 percent of the companies had an active YouTube channel, which was mainly focused on showcasing the organization culture and values. Across the 50 companies, the number of Facebook followers ranged from a minimum of 80 to approx. 29.5 million, Instagram followers ranged from 106 to 2.36 million, Twitter followers from 341 to 4.6 million and LinkedIn followers ranged from 1836 to approximately 18.1 million.

The SMP scores ranged from 5 (Murphy USA) to 46 (Tiffany & Co). OEB scores ranged from 1 (Celgene) to 15 (Bajaj Finserv). Figures 5.3 and 5.4 plot companies' rank along with their SMP score and OEB score, respectively, on the y-axis. The plot shows a negatively sloped trend line in both the cases, implying that companies with higher ranks on the Forbes best employer list performed better with regard to their overall social media presence across the six social networking platforms as well with regard to the presence of employer branding value propositions in their online branding strategy. Thus, companies higher on the list (Forbes 2019) had a better deployment of its SNSs in creating an effective employer branding.

EMPLOYER VALUE PROPOSITIONS

Using thematic analysis (Bazeley & Jackson, 2013), the presence of a companies' online employer brand message as well as the variety of employer branding posts,

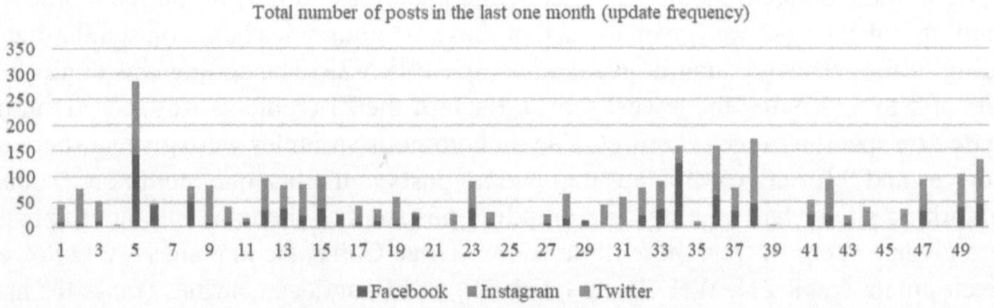

FIGURE 5.2 Update frequency on Facebook, Twitter, and Instagram for each of the 50 companies.

(Source: Authors' analysis using Excel.)

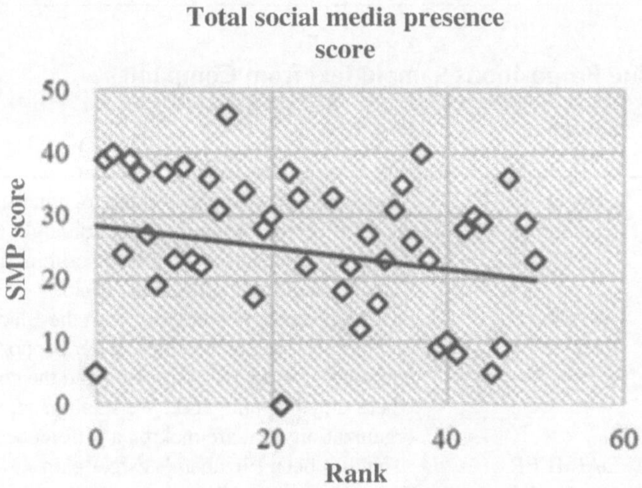

FIGURE 5.3 A trend line of company's social media presence and online employer branding score with rank.

(Source: Authors' analysis using Excel.)

FIGURE 5.4

spread across the five social media platforms were assessed on the eight dimensions of the "Employer Branding Value Proposition" – application value, development value, interest value, social value, economic value, work–life balance management value and brand image (Reis & Braga, 2016; Dabirian et al., 2019). Some examples of the same are presented in Table 5.3.

TABLE 5.3
Employer Value Propositions Sample Text from Companies

Value projected	Source	Post
Application value	CISCO Facebook	"Today and every day, we stand together in support of the LGBTQ+ community to join the fight for equality, inclusion and stand up against discrimination. #WeAre Cisco#CiscoPride#GlobalPride2020."
	AMAZON Instagram	"Amazon stands in solidarity with the Black community. We are steadfast in our support for our employees, partners, customers, and the communities where they live and work. We stand in support of organizations that are making a difference."
	DAIMLER Facebook	"That has been Elizabeth's career path so far! Today, she advises Daimler's developers on how automated driving systems can satisfy our strict ethical standards."
	TIFFANY & CO Instagram	"We are excited to offer volunteer opportunities and career mentorships to the organization's clients. By lending financial support and leveraging our voice, we hope to inspire, empower and create positive change. #TiffanyProud."
Interest value	NESTE Facebook	"I get to be in the frontline of developing our sustainable business model further." Check out our story for a look at how our innovators are creating a healthier planet for our children."
	INTUIT Facebook	"Every day, we're inspired by software engineers like Maya Bello who are committed to shaping the world and helping others through technology."
	MICROSOFT Facebook	"The jitters, the sweat, and long nights, it's all worth the thrill of landing your dream job. Three millennials share how they landed their dream jobs at Microsoft when they took a detour from conventional job hunting."
Social value	SIEMENS Facebook	"Let's bring down stereotypes that only prevent us from connecting on a higher level with our teams, colleagues and friends."
	DAIMLER Facebook	"I love my job because everyone here shares the same vision and is dedicated to work towards it. With our passion, we create a family environment in which we support each other."
	AMERICAN EXPRESS Twitter	"From intern to full-time, #TeamAmex colleague Daniela A. has seen first-hand how Amex embraces diversity. We're proud to back colleagues like Daniela and foster an environment where everyone feels they belong. #HappyPride."

TABLE 5.3
Cont.

Value projected	Source	Post
	RED HAT Twitter	"The best part about being a Red Hatter is the feeling of connection with colleagues across the globe."
Economic Value	SAP Facebook	"#SAP and @girlscouts USA developed a partnership around the concept of equity and fair pay and equal pay."
Development value	IBM Facebook	"Suparna Menon exemplifies IBMers who own their career at IBM. She pivoted from a marketing & strategy background and currently leads the Design practice at IBM iX. Listen in as she shares her career journey and what it takes to make it big in the technology domain. #InspiringLeaders AtIBM."
	DAIMLER Facebook	"From archaeology, human-computer interaction, and futures research to ethics in the Strategy & Stakeholder Management department at Daimler. That has been Elizabeth's career path so far."
Management value	IBM Facebook	"Be passionate and believe in yourself were two valuable tips shared by my mentor with me around two decades back. They have been key to steering my confidence as my career went through its twists and turns."
	VOLKSWAGON GROUP Twitter	"Be passionate and believe in yourself were two valuable tips shared by my mentor with me around two decades back. They have been key to steering my confidence as my career went through its twists and turns." #WeForGreenDeal: CEO-Initiative.
Work/Life balance	RED HAT Twitter	"Every day I'm proud to be a Red Hatter, but today I am especially proud. Thank you to all of the leaders from every level of the company that showed up today. Take more risk. Be fearless. Celebrate achievements. You belong. #Pride".
	SAP Instagram	"Keep yourself healthy, mentally as well as physically while working from home. Take out time from your day to perform #yoga and ensure your overall well-being."
	RED HAT Twitter	"Mom dad and baby #3 are doing great! Glad that @ RedHatprovided parental leave! I feel very fortunate that I get time off during this pandemic to be with my new born. I hope everyone is staying safe."

(*Continued*)

TABLE 5.3
Cont.

Value projected	Source	Post
Brand Image	RED HAT Twitter	"Red Hat believes in the importance of diversity & inclusion, and one of the first steps is making sure everyone feels welcome in the #opensource community. I'm proud of the steps we're taking to eradicate problematic language."
	COCA COLA Facebook	"We will do our part to listen, learn and act. Coca-Cola is committed to making a difference by rallying the strength of our employees, families and friends."
	BMW GROUP Facebook	"We at the BMW Group love and appreciate the #diversity amongst our colleagues and take this as an advantage."
	WAYFAIR Instagram	"We've been listening. We've been learning. We've been reflecting on our organization and refocusing our commitment to diversity – because we believe racism is unacceptable."

Source: Author's analysis.

CORRESPONDENCE ANALYSIS

The companies were divided into five clusters of ten each (top 10, next 10 … last 10), on the basis of their ranks. A correspondence analysis (Hoffman et al., 1986; Greenacre et al., 2007) was performed using SPSS on cluster rank and the eight values (see Figure 5.5). Correspondence analysis gives a spatial presentation by embedding the row and column points of two multinomial variables in a Euclidean space.

The variable plot shows that management value is the least differentiating factor among the clusters as it is closest to the centroid, implying that this theme had an equal weightage (on an average) across all company's SNSs. Cluster 3 and Cluster 4 are the most differentiated on the basis of value type reflected by their SNSs. They have maximum contribution towards the inertia of dimension 1, owing to the huge distance between them as well their placement on the opposite sides of the centroid. Cluster 3 has a stronger association with Interest value vis a vis cluster 5. On the other hand, cluster 4 has a strong negative association with interest vale. Clusters 3 and 5 have more similar value patterns between them as compared to clusters 1 and 2. Cluster 1 is more closely associated with work–life value and social value, whereas cluster 2 is more closely associated with management value. Work–life balance value is majorly represented by clusters 1 and 2. Social value propositions are least represented by clusters 3, 4 and 5.

The number of employer branding posts on each of these platforms and the five criterions that were used to segregate them were coded cluster-wise to give a score

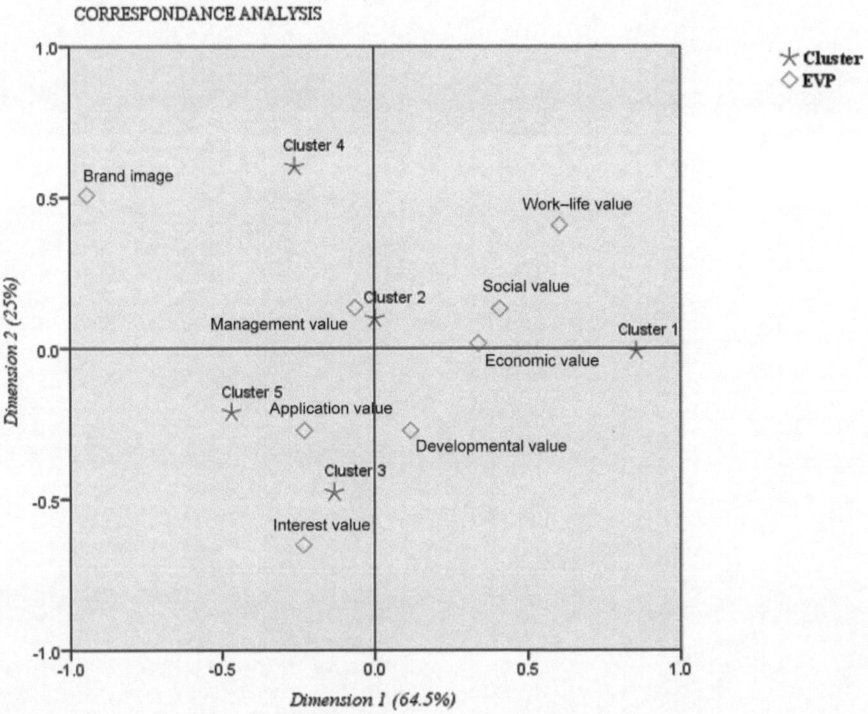

FIGURE 5.5 Correspondence analysis.
(Source: Authors' analysis using SPSS V25.)

out of 5 on each of these dimensions (5 being the highest). Figure 5.6 represents the same in the form of radar diagrams (Gehrels et al., 2016), across all social networking platforms on the basis of clusters (C1, C2, C3, C4 and C5). The corner points represent scoring on the number of employer branding posts as well as the pre-selected five thematic criterions.

Some examples of the posts fulfilling the criterions are as follows: "As a result of passionate employees and desire to support our LGBTQ community, we have introduced new transgender employee benefits including family expansion and medical procedures" (Intuit employee well-being), "Apply now to become a merchant for digital management at Daimler! What is this apprenticeship about? And what are the core fields of learning? Check out our Instagram story to find out." (Daimler-job search functionality), "Kellie and her team have designed a best in market App experiences leveraging native capabilities." (Wayfair- Employee achievements) and "Our intern feels like she is truly living in open culture through collaboration, innovation, and support. Even remotely, I still experience all the Red Hat values: freedom, courage, commitment, and accountability" (Red Hat description of organization culture & values).

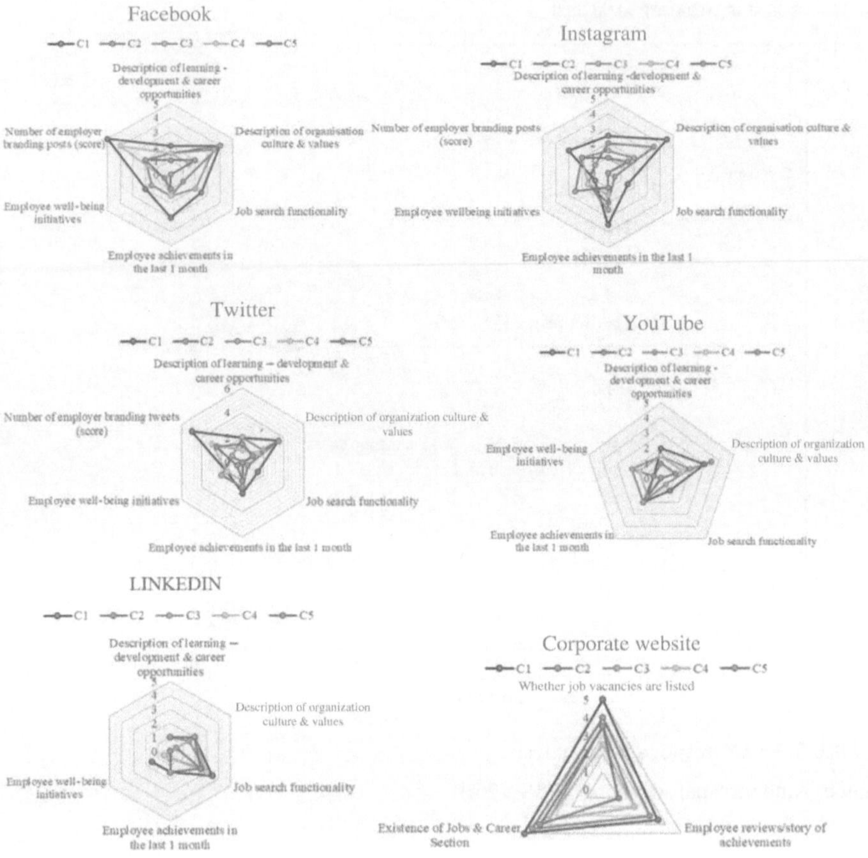

FIGURE 5.6 Cluster-wise employer branding presentation using social networking platforms data. Note: C1, C2, C3, C4 C5 are the five clusters of the companies.

(Source: Authors' analysis.)

DISCUSSION AND MANAGERIAL IMPLICATIONS

The study provides valuable insights into online employer branding strategies of the world's best employers (Forbes, 2019). It presents a quantitative as well as a qualitative overview of the company's social networking platforms and its ability to utilize this valuable resource in projecting the eight employee value propositions that are discussed above. It provides a comprehensive picture to the HR managers on developing an effective employer branding strategy using a costless and a widespread resource – social media.

Employer branding can be perceived as an integrated recruitment strategy that positions an organization in an appealing and an attractive way in the minds of the potential candidates as well as can be instrumental in employee retention and reduced turnover rates (Deal et al., 2010; Tanwar & Prasad, 2016; Rana et al., 2019). A major portion of the "potential employees" category will be formed by the millennials in the near future, who are well equipped and habitual to the use of internet (Schawbel et al., 2015). This generation is characterized by different expectations than the previous generations (Hauw et al., 2010), implying the companies cannot continue to use the old school strategies anymore (Cennamo et al., 2008). This warrants an in-depth strategy by HR managers to use the social media platforms available extensively and optimally at their disposal in establishing their competitive advantage (Schlager et al., 2011; Robertson & Khatibi, 2013).

The company-wise analysis showed better deployment of the company controlled social networking platforms for the companies that were ranked higher on the list. This has direct implications for business leaders across all sectors that social networking tools can be used as a great instrument to attract potential candidates in a targeted manner, strengthen relationships with the current employees, present themselves as a sought-after employer brand, and maintain a pool of highly committed human resource talents. The current practices shown in this research are beneficial for other stakeholders the company as well-both internal and external. Not all the companies were found to be active on all the social networking platforms that were considered in this study.

Managers can use findings of this study to compare their company's social media strategy with the world's top employers and thus make necessary amends to beat the competition (Dabirian et al., 2019). The work–life value, social value and management value were the most highlighted by top companies. Paying attention to these value propositions more, might produce fruitful results for companies' adopting social networking platforms for creating an employer brand. The non-monetary values are more important than monetary values (Sivertzen et al., 2013). The multi-platform mixed analysis approach gives companies' managers easy ideas for increasing their social networking presence effectively.

CONCLUSION AND FUTURE RESEARCH RECOMMENDATIONS

With the advent of social media platforms, new space is opened to engage with people and the impact is also clearly seen in the changing HR practices. This new mode of audio-visual communication, reviews, and comments gave a new meaning to the employer brand attractiveness. This study is unique in providing vast social network platforms content coverage across 50 most attractive employers. The focus on employer brand value propositions is the key to attract and retain employees in these Internet-driven times (Sivertzen et al., 2013). The use of social media as a human resource management instrument as well as an employer branding tool is an effective strategy to build reputation and to create an edge over competitors (Backhaus & Tikoo, 2004; Sivertzen et al., 2013). The chapter only provides a snapshot as it is conducted over a short span of time of one month over which data were collected on a number of dimensions. Further longitudinal studies should be conducted over a longer period of time for in-depth analysis and employer branding strategy trends, for example- trends during recruitment season in particular. Studies can also be conducted for a better understanding of the effect of social media employer branding strategies on the current employees as well as potential applicants via in-depth interviews or focus groups. This research considered top employers in the world. Future research could be carried out on Indian top employers or a comparison can be made between top Indian and world's best employers. A qualitative analysis can be undertaken for knowing the employer's perspectives on companies' value propositions. Building a strong employer brand is a trigger for company's success and has potential for minimizing costs.

REFERENCES

Ambler, T., & Barrow, S. (1996). The employer brands. *Journal of Brand Management*, *4*(3), 185–206.

App, S., Merk, J., & Büttgen, M. (2012). Employer branding: Sustainable HRM as a competitive advantage in the market for high-quality employees. *Management Revue*, 23(3), 262–278.

Arachchige, B. J., & Robertson, A. (2011). Business student perceptions of a preferred employer: A study identifying determinants of employer branding. *IUP Journal of Brand Management*, *8*(3), 25-46.

Backhaus, K., & Tikoo, S. (2004). Conceptualizing and researching employer branding. *Career Development International*, *9*(5), 501–517.

Barney, J. (1991). Firm resources and sustained competitive advantage. *Journal of Management*, *17*(1), 99–120.

Barrow, S., & Mosley, R. (2011). *The employer brand: Bringing the best of brand management to people at work. Global Business Review*, 12(2), 353–354.

Bazeley, P., & Jackson, K. (Eds.). (2013). *Qualitative data analysis with NVivo*. Sage Publications.

Berthon, P., Ewing, M., & Hah, L. L. (2005). Captivating company: Dimensions of attractiveness in employer branding. *International Journal of Advertising*, 24(2), 151–172.

Bhatnagar, J., & Srivastava, P. (2008). Strategy for staffing: Employer branding & person organization fit. *Indian Journal of Industrial Relations*, 44(1), 35-48.

Carpentier, M., Van Hoye, G., Stockman, S., Schollaert, E., Van Theemsche, B., & Jacobs, G. (2017). Recruiting nurses through social media: Effects on employer brand and attractiveness. *Journal of Advanced Nursing*, *73*(11), 2696–2708.

Cascio, W. F. (2014). Leveraging employer branding, performance management and human resource development to enhance employee retention. *Human Resource Development International*, *17*(2), 121–128.

Cennamo, L., & Gardner, D. (2008). Generational differences in work values, outcomes and person-organization values fit. *Journal of Managerial Psychology*, 23, 891–906.

Chatman, J. A. (1989). Improving interactional organizational research: A model of person-organization fit. *Academy of Management Review*, *14*(3), 333–349.

Clair, A. (2016). Employer branding: The role of social media in attracting and retaining talent, a study of Indian IT companies. *Business Dimensions*, *3*(8), 93–101.

Collins, C., & Kanar, A. (2013). Employer brand equity and recruitment research. In: Yu, K; Cable, D. (Eds.), *The Oxford handbook of recruitment*. Oxford Library of Psychology. New York: Oxford University Press.

Dabirian, A., Paschen, J., & Kietzmann, J. (2019). Employer branding: Understanding employer attractiveness of IT companies. *IT Professional*, *21*(1), 82–89.

Deal, J. J., Altman, D. G., & Rogelberg, S. G. (2010). Millennials at work: What we know and what we need to do (if anything). *Journal of Business and Psychology*, *25*(2), 191–199.

De Hauw, S., & De Vos, A. (2010). Millennials' career perspective and psychological contract expectations: Does the recession lead to lowered expectations? *Journal of Business and Psychology*, *25*(2), 293–302.

De Stobbeleir, K. E., De Clippeleer, I., Caniëls, M. C., Goedertier, F., Deprez, J., De Vos, A., & Buyens, D. (2018). The inside effects of a strong external employer brand: How external perceptions can influence organizational absenteeism rates. *The International Journal of Human Resource Management*, *29*(13), 2106–2136.

Doey, L., & Kurta, J. (2011). Correspondence analysis applied to psychological research. *Tutorials in Quantitative Methods for Psychology*, *7*(1), 5–14.

Gehrels, S., Wienen, N., & Mendes, J. (2016). Comparing hotels' employer brand effectiveness through social media and websites. *Research in Hospitality Management*, *6*(2), 163–170

Greenacre, M. (2007). *Correspondence analysis in practice,* Second Edition. London: Chapman & Hall/CRC.

Grzesiuk, K., & Wawer, M. (2018, September). Employer branding through social media: The case of largest Polish companies. In: *10th International Scientific Conference "Business and Management 2018*. Vilnius Gediminas Technical University, Lithuania.

Hoffman, D. L., & Franke, G. R. (1986). Correspondence analysis: Graphical representation of categorical data in marketing research. *Journal of Marketing Research*, *23*(3), 213–227.

Horn, I. S., Taros, T., Dirkes, S., Hüer, L., Rose, M., Tietmeyer, R., & Constantinides, E. (2015). Business reputation and social media: A primer on threats and responses. *Journal of Direct, Data and Digital Marketing Practice*, *16*(3), 193–208.

Kaplan, A. M., & Haenlein, M. (2010). Users of the world, unite! The challenges and opportunities of social media. *Business Horizons*, *53*(1), 59–68.

Kashive, N., Khanna, V., & Bharthi, M. (2020). Employer branding through crowdsourcing: Understanding the sentiments of employees. *Journal of Indian Business Research,* 12(1), 93–111.

Katiyar, V., & Saini, G. K. (2016). Impact of social media activities on employer brand equity and intention to apply. *NMIMS Management Review*, *28*, 11–31.

Kaur, T. (2013). Role of social media in building image of an organization as a great place to work. *ASBBS Proceedings*, *20*(1), 546.

Kissel, P., & Büttgen, M. (2015). Using social media to communicate employer brand identity: The impact on corporate image and employer attractiveness. *Journal of Brand Management*, *22*(9), 755–777.

Lai, L. S., & To, W. M. (2015). Content analysis of social media: A grounded theory approach. *Journal of Electronic Commerce Research*, *16*(2), 138.

Lievens, F., Van Hoye, G., & Anseel, F. (2007). Organizational identity and employer image: Towards a unifying framework. *British Journal of Management*, *18*, S45–S59.

Mičík, M., & Mičudová, K. (2018). Employer brand building: Using social media and career websites to attract generation Y. *Economics & Sociology*, *11*(3), 171–189.

Nagendra, A. (2014). Paradigm shift in HR practices on employee life cycle due to influence of social media. *Procedia Economics and Finance*, *11*(14), 197–207.

Rana, G., Sharma, R., Singh, S., & Jain, V. (2019). Impact of employer branding on job engagement and organizational commitment in Indian IT sector. *International Journal of Risk and Contingency Management (IJRCM)*, *8*(3), 1–17. doi:10.4018/IJRCM.2019070101

Reis, G. G., & Braga, B. M. (2016). Employer attractiveness from a generational perspective: Implications for employer branding. *Revista de Administração (São Paulo)*, *51*(1), 103–116.

Riffe, D., Lacy, S., Fico, F., & Watson, B. (2019). *Analyzing media messages: Using quantitative content analysis in research*. Routledge.

Robertson, A., & Khatibi, A. (2013). The influence of employer branding on productivity-related outcomes of an organization. *IUP Journal of Brand Management, 10*(3), 17.

Robertson, J., Lord Ferguson, S., Eriksson, T., & Näppä, A. (2019). The brand personality dimensions of business-to-business firms: A content analysis of employer reviews on social media. *Journal of Business-to-Business Marketing*, *26*(2), 109–124.

Roy, S. K. (2008). Identifying the dimensions of attractiveness of an employer brand in the Indian context. *South Asian Journal of Management*, *15*(4), 110–130.

Saini, G. K. (2020). Shoppers stop: Leveraging social media for employer branding. *Emerging Economies Cases Journal*, 2(1), 54–61.

Schawbel, D. (2015). How millennials see meetings differently. Available at: www.apbspeakers.com/resources/whitepapers/SchawbelWhitePaper.pdf

Schlager, T., Bodderas, M., Maas, P., & Cachelin, J. L. (2011). The influence of the employer brand on employee attitudes relevant for service branding: An empirical investigation. *Journal of Services Marketing*, 25, 497–508.

Schreier, M. (2012). *Qualitative content analysis in practice*. California: Sage Publications.

Sharma, R., Singh, S. P., & Rana, G. (2019). Employer branding analytics and retention strategies for sustainable growth of organizations. In: Chahal H., Jyoti J., & Wirtz J. (Eds.), *Understanding the role of business analytics*. Springer, Singapore. https://doi.org/10.1007/978-981-13-1334-9_10.

Sirgy, M. J. (1982). Self-concept in consumer behaviour: A critical review. *Journal of Consumer Research*, *9*(3), 287–300.

Sivertzen, A., Nilsen, E., & Olafsen, A. (2013). Employer branding: Employer attractiveness and the use of social media. *Journal of Product & Brand Management*, *22*(7), 473–483.

Tanwar, K., & Kumar, A. (2019). Employer brand, person-organization fit and employer of choice. *Personnel Review, 48*(3), 799–823.

Tanwar, K., & Prasad, A. (2016). Exploring the relationship between employer branding and employee retention. *Global Business Review*, *17*(3_suppl), 186S–206S.

Vaismoradi, M., Jones, J., Turunen, H., et al. (2016). Theme development in qualitative content analysis and thematic analysis. *Journal of Nursing Education and Practice, 6*(5), 100–110.

Weber, R. P. (1990). *Sage University paper series on quantitative applications in social sciences, No. 07-049.Basic content analysis (*2nd *ed.)*. Sage Publications.

Zhang, Y., & Wildemuth, B. M. (2009). Qualitative analysis of content. In: Wildemuth, B. (ed.), *Applications of social research methods to questions in information and library science*. Westport, CT: Libraries Unlimited.

APPENDIX 5.1
Social Media Data Synopsis

SR no.	Information collected with the respect to the following 21 dimensions
1	Facebook Page Likes
2	Average likes on last five Facebook posts
3	Update frequency – number of Facebook posts in the last one month
4	Number of employer branding posts on Facebook (one month)
5	Instagram Page Followers
6	Average likes on last five Instagram posts
7	Update frequency – number of Instagram posts in the last one month
8	Number of employer branding posts on Instagram (one month)
9	Number of Twitter followers
10	Average likes on last five Tweets
11	Update frequency-Number of Tweets in last one month
12	Number of employer branding Tweets (one month)
13	Existence of a YouTube channel (Y/N)
14	Number of YouTube channel subscribers
15	Update frequency-Number of YouTube videos in the last one month
16	Average views on last five YouTube videos
17	Number of LinkedIn page followers
18	Average likes on last five LinkedIn posts
19	Presence of jobs and career section page on corporate website (Y/N)
20	Job vacancies listed on corporate website? (Y/N)
21	Employee reviews/achievement stories on corporate website (Y/N)

6 Corporate Social Responsibility to Corporate Environment Ready

A Paradigm Shift to Organizational Branding

Chandan Veer and Pavnesh Kumar

INTRODUCTION

The business organizations dominate the society and environment, that is, they shape, nurture, and control the society. Any change in the society is partially or fully influenced by the decisions of these business organizations. Globalization has thrown many new challenges among the competing organizations, such as environmental responsibility, healthy financial position, happy and satisfied customers, cost minimization, and sustainable practices. To mitigate these challenges, globalization has given birth to the practice of the concept of corporate social responsibility (CSR). Gradually and day by day, the scope of CSR is expanding. In the initial days of CSR, welfare of employees was the core of CSR. After a certain period the society also got added in the horizon of CSR and then philanthropy came into practice. Even today many business houses are still working on this particular concept. Now it is the call of time to rethink on the concept and practices of CSR. The organizations will have to broaden the purview of CSR which should include both internal and external environments as a whole.

CSR is an ethical practice of organizations to contribute to the growth and development of employees, stakeholders, local community, and society, with a view to gain organizational reputation.

CSR is a business technique and when executed properly it facilitates brand building of the organization. It is always expected from the organizations to think beyond financial benefits and try to consider the social, environmental, and ecological impacts of any business decisions.

The World Bank describes CSR as

> [T]he commitment of business to contribute to sustainable economic development, working with employees, their families, the local community, and society at large to improve their quality of life, in ways that are both good for business & good for development.
>
> (Public sector roles in strengthening corporate social responsibility: taking stock, H Ward, 2004, p. 9)

Corona crisis has given a new challenge to the corporate world in terms of its business functions and survival. The time has come that the business houses act more in terms of health and the environment. The corporate house associated with any industry and doing business will have to rethink in terms of its people, process, planet, and profit. The time has come to reinvent the meaning of CSR. In this chapter, we will try to frame the activities of CSR in a broader sense with the help of a model termed as corporate environment ready (CER) model. This model will be discussed in detail in the further section of this chapter. However, before that let's have a conceptual understanding about the CER model. The term CER is self-explanatory with its word.

CER is an integrated approach of an organization engaged in business activities, which is environmentally, socially, and ecologically favorable for both the organization and society. It is a win-win strategy adopted by an organization with an aim to have sustainable growth and development. CER is a long term strategy of an organization that aims to contribute to the employees, stakeholders, consumers, society and environment safety, and growth and development. The CER model can give leverage to the company's growth and development along with a strong and stable organizational reputation.

Organizational branding is the process of expressing or communicating to the community about the symbolic organizational mission and the prevailed organizational practices. The smaller the difference between symbolic organizational mission and the prevailed organizational practices, the higher will be the organizational branding.

Organizational branding is an integrated marketing communication technique, which tries to create a positive brand image in front of the society.

Many companies are practicing from a long time some sort of social and environmental activities with a narrow goal, that is, to contribute to the prosperity of the society on which they depend for various kinds of resources. The companies basically practice CSR to defend themselves for the wrong deed. But now the current situation has raised a question or created a pressure to reframe CSR as a business discipline. Many corporate giants perceive that CSR stymies the main business goal. The CER model may be found to be impressive and making sense to those who perceive CSR as an obligation rather than a responsibility. There is always a difference in opinion among the business houses that the business goals and motives cannot be aligned with the company's social and environmental activities. But what if a new reframed CSR, that is, CER activities, reduces business risk, enhances reputation, and contribute in business goals, then it is all to the good. The CER model can lead to sustainability, triple bottom line, blue ocean strategy, competitive advantage, and strong brand equity. This chapter will try to explain why firms should reconsider their CSR activities and provide a systematic process in CSR to undertake it in the form of discipline of business organizations.

LITERATURE REVIEW: A BRIEF STORY

The concept of CSR is existing in history but it got attention and recognition only in the last decade. In the 18th century, employers understood the importance of an efficient human resource in an organization and that a lack of proper food, medical facility, and housing will hamper the productivity of the workforce which will ultimately impact organizational growth. In recent times, medical, housing, and subsidized food facility is seen as a philanthropic effort, but in the past it was due to the self-interest of manufacturers (Carroll & Shabana, 2010).

Bernard Dempsey in 1949 in his article, "The Roots of Business Responsibility," published in *Harvard Business Review*, suggested a guideline for responsible business practices. He suggested four concepts in terms of justice: Exchange Justice – the trust lying in the exchanges in the market; Distributive Justice – the trust between government and people; General Justice – acceptance of legal framework along with ethical obligations; and lastly, Social Justice – the responsibility to contribute in the well-being and growth of individual and community.

Two months after Dempsey's article, Donald K. David argued upon business houses to contribute to public well-being and think beyond immediate business activities. Dempsey and David both asserted the reason behind the social obligation. (1) If there will be no man, then no business will even exist. (2) A well functional society will contribute in the excellence of the operations of the business. (3) The business houses have a control on resources and they have the capability to improve the status of society.

Bowen (1953 defined CSR as an obligation, decision, course of action, repercussions or policies which lead to the desirable objective and values for the society.

Heald (1971) expressed in his article which was actually a comprehensive history from 1900 to 1960, whose topic was "The social responsibilities of business: Company and Community 1900–1960." He mainly focused on the practices of social responsibility carried by business leaders.

The Committee of Economic Development (CED, 1971) explained CSR as a business house that functions with the public approval and with a basic purpose of serving the society and satisfying their needs.

Eilbert and Parket (1973) explained CSR as to think it like a camaraderie. The concept is expressed in two ways: the first one is to avoid doing anything that will spoil the relationship and the second one is to be committed to help solve the problems of others or the commitment in general to play an active role in the alleviation of social problems.

Sethi (1975) distinguished the terms like "social obligation," "social responsibility," and "social responsiveness." He explained social responsibility as an activity that implies elevating corporate behavior up to that level where it is identical with the prevailing social practices, values, and expectations.

Carroll (2016) propounded a CSR model. This model categorizes the CSR into four different responsibilities, that is, economic, legal, ethical, and philanthropic responsibilities. The economic responsibility suggests the organization to be profitable as all other activities depend on it. Legal responsibility suggests to obey the incorporated legal framework, that is, law. Ethical responsibility suggests doing fair and right practices in the business and avoiding harming the society. The philanthropic responsibility emphasizes being a good corporate citizen.

Jones (1980 cited in Carroll 1999) explained CSR as an activity which is broader in sense, that is, it is not confined to shareholders only but it comprises a society that includes customers, employees, stakeholders, and community. The CSR is a self-driven initiative and not influenced by the law.

Arlow and Gannon (1982) in their research found that high CSR performance leads to higher company performance, and all else being equal. CSR reveals the personal traits of the top management of an organization (Hambrick & Mason, 1984), so the top management engaged in CSR discloses values, preferences, and inclination which actually reduce information asymmetry between the company and society (Bitektine, 2011).

Frederick (1986) expressed that CSR should be connected with business ethics. He emphasized that morally correct business activities will lead to CSR that is, Corporate Social Responsibility.

The World Business Council for Sustainable Development (WBCSD, 1998) defined CSR as a continuous obligation of business houses to act ethically and come up with economic development and improving the living standard of the workforce along with their family and society at large.

Williams and Siegel (2001) defined CSR as an engagement in CSR activities that is beyond government regulations and by doing this an organization can differentiate itself within a group.

Margolis and Walsh (2003) expressed that financial aspects of CSR performance have been extensively researched, but very few studies have been conducted about the other effects of CSR. Wood (2010) and Margolis and Walsh (2003) closely studied the CSR performance and asserted that it is the time to divert the focus from how CSR affects the firm toward how the firm can impact stakeholders and society as a whole.

Mari Kooskora (2005) propounded CSR is how the companies carry their business process to have an overall positive influence on the society.

The International Labour Organization (ILO, 2007) stated that CSR is a strong belief in the principles and values of an organization which considers the impact of business activities on society. It is a self-driven activity carried out to exceed the compliance with the law.

CSR activities have a positive correlation with earnings forecasts (Lee, 2017) and thus open information sharing or communication (Jo & Kim, 2008).

Socially responsible organization attracts more investors (Day, 2001) and thus enables to enhance firm reputation (Jeong et al. 2018).

Blowfield and Murray (2008) asserted that many researches have been done on the impact of CSR on business and its benefits for business but only a few about how CSR affects the major social issues it tried to tackle.

According to Bitektine (2011, p. 160), "the sets of dimensions that from [sic] reputation & legitimacy are often overlapping, and so the same dimensions can be used to make legitimacy and reputation judgments."

Michael Hopkins (2014) defined CSR as a combination of three components: (1) corporate, (2) stakeholders and (3) ethical behavior. Corporate means group of people working together in an organization for profit. Stakeholders comprise customers, employees, suppliers, shareholders, and the local community. Ethical behavior comprises values, code of conduct, and corporate governance.

CSR BY INDIAN COMPANIES: AN OVERVIEW OF SELECTED COMPANIES

We can say Indians have a great interest in CSR. In India CSR is regulated as per "The Indian Companies Act-2013", Schedule VII consisting of 10 areas, giving a way to Indian companies to carry CSR activities. The Indian companies have to spend 2% of their net profit in CSR activities or CSR projects.

In this section, we will try to explore the CSR activities of a few reputed Indian companies on four different subjects: *profit, people, planet* and *process.* In profit section, we will just consider how sound the company is, that is, the financial position of the organization as profit is the base of CSR.

In people section, we will analyze the various CSR activities moving around the people's welfare, development, and safety.

In planet section, we will examine the environmental activities carried by these organizations keeping in view their environmental footprints.

In process section, there will be a different angle of study, as the past researchers have explored about the process with an external aspect. We mean to say that we will look over the manufacturing, distribution, waste management, supply chain management, packaging, etc. The motive behind such exploration is to make organizations understand and believe that these internal business processes influence the CSR activities and organizational reputation.

Tata Group: A conglomerate diversified business enterprise which is in steel, automobile, software, consumer durables, and fashion brands and accessories, the Tata group has more than 80 companies. The group spends more than the expected figure as per the Indian Companies Act of net profit in various CSR activities, exceeding the government expectations. The group has had a very healthy financial positions right from its establishment to the current time. The company's CSR can be broadly classified as promoting education, health, livelihood, and rural and urban infrastructure development. Besides this they undertake areas of sports, disaster relief, environment, and ethnicity with the aim to improve the quality of life of communities.

Reliance Industries Limited: It is one of the leading business houses of India, a conglomerate diversified in IT, oil and petroleum, digital services, media and entertainment. Financially it is a sound company with the highest market capitalization in India as of today. They are highly committed towards the CSR activities under the name of Reliance Foundation and umbrella organization for social initiatives. They have a firm belief in sustainable development, so their initiatives happen to be in conformity with the Sustainable Development Goals (SDGs) of the United Nations. They are focusing on rural development, environment conservation, health, education, sports, art and culture, water security, and many more as the list will go on.

Indian Oil Corporation Limited: It is India's leading oil and petroleum public sector company. The company's sustainability and CSR activities focus on providing safe, efficient, and ethical customer services, which can curb the negative impact on environment and enhance the quality of life of community. They focus on efficiency in operations and processes, their CSR activities include safe drinking water along with water conservation, healthcare, education and employment generating training skills, women empowerment, and environmentally sustainable practices within and beyond the organization.

Suzlon Group: Suzlon Group is one of the world's largest wind turbine manufacturers providing green energy solutions. The Suzlon Group believes that sustainability can be achieved by the integration of all resources, that is, financial, natural, social, and human, which ultimately enhances the business advantages. Suzlon CSR model "SUZTAIN" incorporates the implementing and monitoring of CSR activities. They have a strong belief in 100% green business initiatives. They are involved in livelihood enhancement, education, environment, health care, and civic amenities.

Wipro: Wipro limited is a giant global information technology consultant and business services providing company. The company is always recognized for its philanthropic commitment, sustainability, and as a good corporate citizen. The domain of CSR activities includes education, healthcare, community ecology, business sustainability, and faculty development programs in various engineering colleges across the country.

SBI: India's largest commercial bank decorated with a crown of premier institution in financial sector. This group has its footprints across several countries in the world. The bank firmly believes in its business sustainability and offers products and services in a responsible manner taking the social and environmental aspects in to account. The bank's digitalization commitment and its green banking services have reduced its carbon footprints and get it closer to SDGs. Paper usage reduction, solar ATMs, community development programs, and health care initiatives show their commitment toward the society and environment.

ITC: ITC is one of India's leading private sector companies with magnificent market capitalization. It has presence in FMCG, hotels, agri business, packaging and paper boards, and specialty papers. The company is carbon positive for many years. The company is engaged in afforestation programs, water shed development, livestock development, education, health and sanitation, solid waste management, and organic farming. ITC carries its CSR activities through public–private partnership as they consider government as an important stakeholder in terms of CSR implementations.

Coal India Limited: An organization committed to fulfill the energy requirements of the country in the primary energy sector. Coal India uses CSR as a strategic tool for sustainable development. They believe to integrate business process with social processes. The areas which they cover under their CSR policy is poverty eradication, hunger and malnutrition, health care, education, vocational skills development, air, water and soil conservation, protection of flora and fauna, and rural development projects. They carry their CSR activities with the local government bodies. They encourage the development and introduction of environment-friendly technologies.

Hindustan Unilever: Hindustan Unilever (HUL) is India's largest fast-moving consumer goods company working for the last 80 years in India. A company with a large brand portfolio is a part of everyday life of millions of consumers. The company as per schedule VII of Indian Companies Act carries various CSR activities like Project Shakti – to promote education, Swachh Aadat and Swachh Bharat – to eradicate hunger, poverty, promotion of sanitation and health care, promotes ecological balance, protection of animals and forest, promoting education, and disaster relief programs. The various programs of HUL have been integrated with their business, that is, CSR and business processes moves parallel.

OBJECTIVES OF THE STUDY

The objectives of this research study are as follows:

- To examine the scope of CSR in building the organizational reputation.
- To understand the limitations of CSR.
- To suggest CER (Corporate Environment Ready) model for building organizational reputation.
- To suggest to implement the CER (Corporate Environment Ready) model as business strategy tool.

RESEARCH METHODOLOGY

Exhaustive literature review regarding the title and its related concepts was done. Secondary data inclusive of quantitative and qualitative were analyzed. The sources of information are from various research publications, published newspapers, journals – online and printed, books, magazines, and websites. The information was collected from libraries and websites. The literature was cross-checked and validated to give the latest information.

CORPORATE ENVIRONMENT READY MODEL

Before discussing the CER model, let's have an overview of some models which explain the scope of CSR for companies. These models actually suggest a way to move toward social responsibility or obligation of an organization toward the society and environment. A few important and notable models include Carroll's model, Halal's model, and Ackerman's model.

Carroll's Model

Archie B. Carroll in 1979 propounded his hierarchical model defined as CSR for the business houses toward the society. His model is in pyramid shape having four categories, starting from the bottom to the top. The four categories are:

i. Economic: This category is at the bottom of the model, mainly expressing about profit for the organization.
ii. Legal: It explains about compliance as per law.
iii. Ethical: It explains about the norms or practices which the society expects from the business organization.
iv. Philanthropic: It explains about the engagement of business enterprises in community development or other social projects.

Halal's Model

This model discusses two aspects, the first one is about the organizational work, interest, and distribution of profit. The second aspect is about social activities. He suggests to have a coordination between the organization and social decisions so that the future of the organization and society would be safe.

Ackerman's Model

This model has three phases for defining the CSR.

i. First Phase: To recognize the social issues to be addressed by the top management.
ii. Second Phase: To appoint an employee or specialist to understand the issues and take measures to eradicate such issues.
iii. Third Phase: To implement the strategy suggested by the specialist.

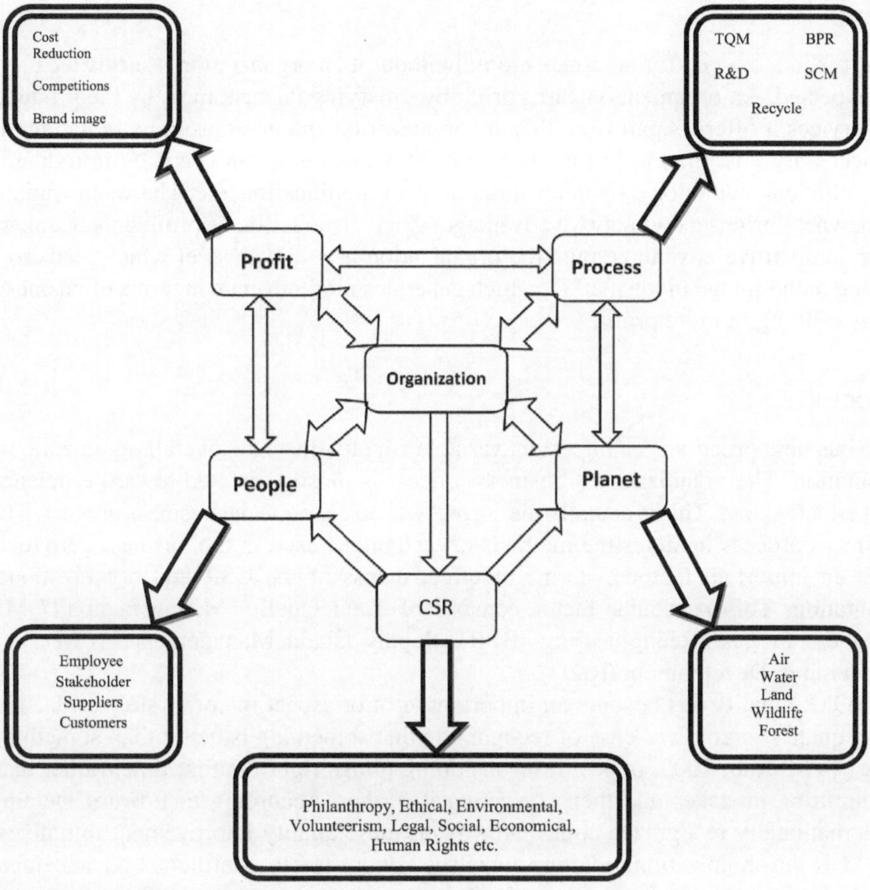

FIGURE 6.1 Corporate environment ready model.

(Source: Authors' own work.)

THE ELEMENTS OF THE CER MODEL

Today brand image of an organization is one of the most important factors that is considered while evaluating the reputation of any organization. This reputation depends on many factors which are directly related to the organization, such as profit earned, employee's welfare, business process, environment friendly, and CSR. The CER model is an integrated model or coordination of the abovementioned factors which contributes to the enhancement of organizational reputation (Figure 6.1).

Let's understand what the model tries to explain.

An organization is a body which is interconnected with profit, process, people and planet. The organization cannot survive in the absence of any one of these factors. Let us explain it in detail successively and individually.

Profit

Profit is like oxygen for an organization; without it an organization's survival cannot be expected. An organization earns profit by satisfying its customers by the products or services it offers. Apart from this the organization enhances profit by reducing the unnecessary cost incurred in the business. This reduction can occur by introducing new efficient technology, by innovation, product modification, etc. These innovations or newness brings a competitive advantage which also facilitates profit enhancement. The competitive advantage puts the organization at a higher level which leads to a strong brand image in the market, which generates more revenue in terms of business, thus resulting in more profit.

Process

The business process is an important variable which affects the overall organizational reputation. The organization's business process is mostly directed toward efficiency and effectiveness. This is actually the correct way to examine the business process. The business process includes the methods or techniques used in the business activities. It is an important factor in terms of effectiveness of the CSR and organizational reputation. This particular factor consists of Total Quality Management (TQM), Business Process Reengineering (BPR), Supply Chain Management (SCM), and Research & Development (R&D).

TQM: Quality has become an important term or aspect in corporate world. "The term quality refers to a sense of recognizing that something is better than something else" (VSP Rao, 2005, p. 60). It means doing things right the first time, rather than committing mistakes and then correcting it further. According to Edward Deming (internationally recognized contributor to Japanese quality improvement initiatives) TQM is an organizational culture which is committed to continuous advancement of skills, process, product and services, and customer satisfaction. TQM is termed as an organizational culture because it is seen as a deeply inherited practice which reflects in every aspect of the organization. A few of the Deming's advice on quality achievement are the following:

a) Always have a reason or purpose for improvement of process and cost reduction.
b) Stop recognizing business on the basis of price tag only.
c) Constantly and continuously improve the production system and services.
d) Accept modern methods of technology.
e) Create a sense of innovation in the organization.
f) Take actions to achieve positive transformation.

The main idea: TQM

i. Do it right in the first time.
ii. Be customer oriented.
iii. Adapt continuous improvement as an important and inseparable aspect of life.
iv. Build a team and empower them.

BPR: Business Process Reengineering primarily focuses on doing things in better way by continuous and incremental improvements. BPR means an evaluation of what an organization is all about, thus making significant changes for the organizational success and for achieving its core values.

R&D: Research and Development – every organization is mostly dependent on the R&D team for some new or improved innovative products or services. The trend which is seen in R&D department is to mostly improve or innovate existing or new products or services on the parameter of cost reduction, usage, added features, packaging and handling, recycling of wastages, etc. The evaluation of the above parameters is necessary, but evaluating them in terms of the societal and environmental impact is also important. This small improvement can add more to the organizational reputation. This new approach can save resources of the organization which can be utilized for more beneficial activities.

SCM: There is a term in computer language "Garbage in, Garbage out". The supply chain management involves this same concept. It involves flow of raw materials to the final distribution of products and services. Suppose if an organization is procuring below standard raw material with a view to reduce cost, then the final product after processing will be of a low standard finished good. This will have its own consequences in terms of health and on the environment. The top management will have to consider it too as it also impacts organizational reputation.

People

Corporate is nothing but a group of people engaged in an integrated activity to run the business keeping in view the profit or the expected common end result. The group of people in this model comprise employers, employees, suppliers, stakeholders, and customers. Employees here represent the people working in any organization, with their family members, friends, and relatives who constitute a society or community.

Stakeholders comprise the owners, who have initiated this activity or taken a risk with an objective of earning profit and serving the community.

Suppliers are those members who are facilitating the business activity from outside, right from the raw material procurement to the distribution of finished goods.

The customers are those who actually purchase the products or avail the services to satisfy their needs and wants.

Planet

Planet is a silent player which is contributing in a major manner in the business functioning, with its limited resources without asking a single question. The earth is providing its resources by the composition of air, land, and water along with from the forest and wildlife. The efficient utilization and conservation of these resources is necessary for long-term survival of a business.

CSR and Four Factors

All these interrelated factors play a vital role in the existence of CSR and make CSR effective and meaningful. If you will give an insight, you will find that most of the organizations (excluding the exceptional ones) focus on only three factors of CSR, that is, people, profit, and planet. They either ignore or not consider process as an important factor for success of a CSR activity.

Profit is something that determines the level of CSR activity by an organization. So if there is no profit, then there will be no CSR as it is the base of this activity. So undoubtedly considering profit is necessary.

CSR from one point of view can be said that it is for people, whether be an employee who expects both monetary and non-monetary benefits in return for the contribution toward the organizational growth. Customers and members of the society are the users of the products or services and they also contribute in the success of the business, so they are also part of CSR. Suppliers are those who facilitate the business activities. The stakeholders are the actual players or planners of CSR because of government regulations as well as their self-interest.

The planet is the purveyor of resources, so it becomes the obligation of the business houses to act toward the safeguard and conservation of the planet. But there is a loophole in the practicing of conserving or protecting the planet. Suppose an organization is involved in a business activity and the wastage of the industry is being disposed in the nearby area whether in a river or on land. At the same time the company is involved in CSR activities like education, wildlife protection or ancient monument conservation. Then this good deed can't justify the aforesaid wrong practice. The issues should be resolved at the originating point, that is, water should be treated properly and then discharged to the river. A few organizations are doing this job.

The composition or integration of all these four factors, that is, people, planet, process, and profit, will take CSR to a new level with more effectiveness. Then the organization will be termed as CER (Corporate Environment Ready). This CER label will undoubtedly enhance the reputation of the organization. The CER model can result in various advantages, such as strong brand equity, competitive advantage, reduced cost, delighted customers, motivated employees, and a balanced ecology.

FINDINGS

- CSR is considered as a moral responsibility which a corporate pursues.
- CSR is mostly philanthropic in nature while CER is both sustainable and philanthropic.
- CSR is determined after profit while CER actually strives to enhance profit and a higher level of CSR.
- The organizations are focusing on profit, people, and planet only while planning the CSR objective.
- There is a strong and hidden link between business processes and CSR.
- CSR seems to be a tool to mitigate the wrong deeds of corporate while CER can improve the process which will add more value to the organizational reputation.

SUGGESTIONS

CSR seems to be an orthodox style, which is actually not clear in its objective. So my suggestions for the corporate firms are as follows:

- CER model can be helpful in curbing down the environmental footprints.
- CER model can positively impact the organizational branding.
- CER model will lead to enhanced profit through cost reduction.
- CER model will improve the internal business process.
- CER model will lead to sustainable development.
- CER model will lead to motivated employees, confident stakeholders and a satisfied customer base.

CONCLUSION

To conclude, organizations will have to understand that the CSR is not effective in current scenario due to its philanthropic nature. The organization will have to think beyond the social obligation, and then only sustainable development will come into picture. CSR can reduce poverty, malnutrition, and improve health, but it will be a temporary solution. The actual impact of business firms is on overall environment. So there is a need to reinvent CSR, that is, to think beyond philanthropy and move toward sustainability through CER.

In the end, a company which is philanthropic but is not aware with its broader role will not survive very long. As per Michael Porter, if any company wants to be good by donating money for social cause only, is actually wasting money, that is, not sustainable, then shareholders will lose interest in long run.

REFERENCES

Ackerman, R. W., & Bauer, R. A. (1976). Corporate Social Responsiveness: The Modern Dilemma. Reston, VA: Reston Publishing Company.

Arlow, P., & Gannon, M. J. (1982). Social Responsiveness, corporate structure, and economic performance. *Academy of Management Review*, 7(2), 235–241.

Bitektine, A. (2011). Toward a theory of social judgments of organizations: The case of legitimacy, reputation, and status. *Academy of Management Review*, 53, 151–178.

Blowfield, M., & Murray, A. (2008). *Corporate Responsibility: A Critical Introduction*. Oxford: Oxford University Press, pp. 452.

Bowen, H. (1953). *Social Responsibilities of the Businessman*. New York: Harper and Row.

Carroll, A. B. (1991). The pyramid of corporate social responsibility: Toward the moral management of organizational stakeholders. *Business Horizons*, July/August, pp. 39–48.

Carroll, A. B. (1999). Corporate social responsibility: Evolution of a definitional construct. *Business & Society, 38*(3), 268–295.

Carroll, A. B. (2016). Carroll's pyramid of CSR: Taking another look. *International Journal of Corporate Social Responsibility,* 1, 3. https://doi.org/10.1186/s40991-016-0004-6

Carroll, A. & Shabana, K. (2010). The business case for corporate social responsibility: a review of concepts, research and practice. *International Journal of Management Reviews, 12*. 10.1111/j.1468-2370.2009.00275.x.

Davis, K. (1960). Can business afford to ignore social responsibilities? *California Management Review,* 2(3), 70–76.

Doh, J., & Guay, T. (2006). Corporate social responsibility, public policy, and NGO activism in Europe and the United States: An institutional-stakeholder perspective. *Journal of Management Studies*, 43, 47–73.

Eilbert, H., & Parket, I. R. (1973). The current status of corporate social responsibility. *Business Horizons*, 16, 5–14.

Frederick W. (1986). Available at: www.williamcfrederick.com/articles/GrowingConcern.pdf

Halal, W. E. (2000). Corporate community: A theory of the firm uniting profitability & responsibility. *Strategy & Leadership*, 28(2), 10–16.

Hambrick, D., & Mason, P. (1984). Upper echelons: The organization as a reflection of its top managers. *Academy of Management Review*, 9, 193–206.

Heald, M. (1971). The Social Responsibilities of Business: Company and Community, 1900–1960. Cleveland, OH: Press of Case Western Reserve University. Doi: 10.1086/ ahr / 76.5.1615-a

Hopkins, M. (2014). What is corporate social responsibility all about? *Aspirare*, I. www.researchgate.net/publication/246912286.

Hopkins, M. A. (2003). Planetary Bargain: Corporate Social Responsibility Comes of Age. Macmillan, UK, 1998; updated and re-printed by Earthscan, 2003 and again reprinted by Routledge, UK, 2010.

ILO (2007). Available at: www.ilo.org/wcmsp5/groups/public/---ed_emp/---emp_ent/---multi/documents/instructionalmaterial/wcms_227866.pdf

Jenkins, R. (2005). Globalization, corporate social responsibility and poverty. *International Affairs,* 8I(3), 525–540.

Jeong, K.-H., Jeong, S.-W., Kee, W.-J., & Bae, S.-H. (2018). Permanency of CSR activities and firm value. *Journal of Business Ethics*, 152, 207–223.

Jo, H. & Kim, Y. (2008). Ethics and disclosure: A study of the financial performance of firms in the seasoned equity offerings market. *Journal of Business Ethics,* 80, 855–878.

Kooskora, M. (2005). Perceptions of Corporate Social Responsibility among Estonian Business Organisations. *EBS Review*. 2005. 74–87.

Lee, D. (2017). Corporate social responsibility and management forecast accuracy. *Journal of Business Ethics*, 140, 353–367.

Margolis, J. D., & Walsh, J. P. (2003). Misery loves companies: rethinking social initiatives by business. *Administrative Science Quarterly*, 48, 268–305.

McWilliams, A., & Siegel, D. (2000). Corporate social responsibility and financial performance: Correlation or misspecification? *Strategic Management Journal*, 21, 603–609.

McWilliams, A., & Siegel, D. (2001). Corporate social responsibility: A theory of the firm perspective. *Academy of Management Review*, 26, 117–127.

Rao, V. S. P. (2005). *Human Resource Management.* New Delhi, India: Excel Books, pp. 60–65. ISBN- 81-7446448-4.

Sethi, S. P. (1975). Dimensions of corporate social performance: An analytical framework. *California Management Review*, 17(3), 58–64.

Srivastava, A. K., Negi, G., Mishra, V., Pandey, S., & Murti, S. (2016). Corporate social responsibility: A case study of TATA Group. *IOSR Journal of Business and Management (IOSRJBM),* 3(5), 17–27. www.iosrjournals.org

Stewart, R. M., Eden, L., & Li, D. (2018). University of Texas, A&M University & Kelley School of Business, USA, CSR reputation and firm performance: A dynamic approach. *Journal of Business Ethics.* doi.org/10.1007/s10551-018-4057-1

Wood, D. J. (2010), Measuring Corporate Social Performance: A Review. *International Journal of Management Reviews*, 12, 50–84. https://doi.org/10.1111/j.1468-2370.2009.00274.x

7 Branding through Workforce

Antima Sharma and Rinku Raghuvanshi

INTRODUCTION

To run a fortunate business, it is crucial to build an image of the organization. Building an image is seen as branding. Big business houses make the best efforts to build their brand image among customers, employers, suppliers, etc., to get recognition for their product. Branding is the most important process to differentiate yourself from the competitors and it also gives clarifications that what you are offering is a better choice. Branding helps the organizations to stand out from the crowd in the global market through brand image. Branding is something that helps us to decide which particular brand to buy (for, e.g., which coffee house is good, which brand shoes to wear, which tea to drink, and so on). The brand is something perceived in the minds of people from which they can connect either with customers or employees (Rana, & Sharma, 2019). Organizations always make new strategies to push their products in the market from time to time through internal and external branding. Generally, businesses spend a huge amount on advertisement, logos, taglines, PR, etc. In today's scenario only external branding measures cannot put much impact on the endorsing of the brand, and there is a need for internal branding also. Internal branding means

TABLE 7.1
Differentiation between Product, Brand, and Branding

Particulars	Meaning	Definition	Example
Product	Any goods and services provided by the manufacturer and seller to satisfy need or want.	Broadly, a product is anything that can be offered to a market to satisfy a want or need, including physical goods, services, experiences, events, persons, places, properties, organizations, information, and ideas.	Water, Bread, Cold Drink, TV Channel, etc.
Brand	The Brand is a special feature, that is, name, term, design, symbol, ingredient, etc. which distinguish one seller from another.	"A brand is a reason to choose."	Bisleri, Harvest Gold, Coco-Cola, Aaj Tak
Branding	Branding is the process to make your product as brand or shaping brand in the consumer's mind.	"Branding is endowing products and services with the power of a brand."	Coco-Cola-Bottle, Aaj Tak-Tag Line, Healthy Grains Bread from-Harvest Brown Bread

to ensure that internal resources are aligned with brands and most importantly the human resources. Internal branding is a process which makes employees understand about the brand and can inject attributes of the brand into the employees. Internal branding means to make employees to be a brand ambassador so that they reflect it to the world in their ways. Further in this chapter we will understand about employee branding and its aspects in detail.

BRANDING

Before proceeding further in detail about branding, there is a need to understand the difference between product, brand, and branding (Table 7.1).

Branding in general is a marketing concept where a company makes known its name, symbol, logo, design of product, features of product, etc., that is easily identified by potential customers. Branding is not only about the sign, symbol, design, or logo of the company, it is a bigger term that also defines the company's value, meaning, vision, missions, employees and the way of involving their stakeholders. Branding is a process to make an organization from ordinary to a brand. Organizations keep on making a strategy to branding their product from different techniques to sustain in a competitive market. Branding helps to create:

Brand Recognition: Identity of the product/organization among the interest group. (Maggie=2 Minutes)

Brand Preferences: To bring loyalty to the product among customers. (I purchase always Nokia Phone)

Brand Awareness: Always strike into the minds of the customers when they purchase the same category product.

Brand Association: Linking your brand with celebrities, images, or geographic regions.

All of the above together is called branding to push a product in the market. Big business houses do much effort and adopt innovative techniques for branding (see Figure 7.1).

Figure 7.1 shows some of the ways to build a brand of the product in the market. These are some of the conventional practices followed by the organization to keep their products alive in the competitive world. Moreover, the conventional methods of marketing, in terms of how companies communicate about their product to customers, are challenged and questioned by scholars (Pilotta et al., 2004). Apart from the external branding of products, now organizations are focusing to build a strong foundation branding; therefore, they realize that there is a need for internal branding. Employees may also influence the organization's external branding positively. Mangold and Miles (2007) consequently argue that they can also tarnish the brand image. Internal branding means connecting the brand with people who are part of the organization – they are customers, investors, suppliers, third parties, and most importantly employees. Internal branding can be created through trust, credibility of employees with each other, making them understand the company's mission, more transparency, and engagement of the employees. The only key resource

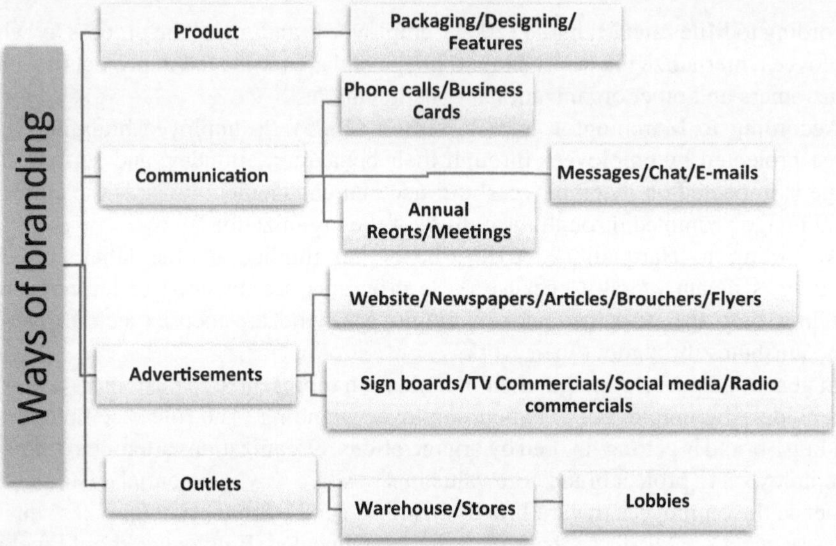

FIGURE 7.1 Ways of branding.
(Source: Authors' own work.)

is an employee that can represent the brand in the best way in the market. For this, there is a need to make employees brand ambassadors or we can say branding through employees/workforce. We will make clear about this by discussing in detail about employee branding or branding through the workforce.

EMPLOYEE BRANDING

There is a new concept "Living the brand" (Ind, 2001; Pringle & Gordon, 2001; Mitchell, 2002 cited in Harquail 2007), which is advertised as strengthening an organization's position in the competitive marketplace while enhancing internal organizational effectiveness, all by emphasizing organization-wide employee involvement in branding processes. The organization's image is represented not only by buildings, assets, financial conditions, and products, but also through its workforce that play a key role in reflecting the image of an organization among the stakeholders. Nowadays, employee branding is a new trend in branding the organization (Sharma, Singh, & Rana, 2019). After understanding these, organizations are gearing up their image through branding their workforce and workplace. Employee branding means transforming your employees into a brand ambassador of the organization so that they can present the company's brand in the best possible way. When employees feel engaged, connected, and satisfied with their work and organizational culture, they show that in their behavior. The process of internally communicating and motivating an employee about the brand is called employee branding and it is described by various scholars in different ways.

DEFINITIONS OF EMPLOYEE BRANDING

According to Miles and Mangold (2004), employee branding is "the process by which employees internalize the desired brand image and are motivated to project the image to customers and other organizational constituents."

According to Marchington and Wilkinson (2005), "[e]mployee branding is the image projected by employees through their behaviors, attitudes, and actions. This image is impacted on by employees' attitude and engagement towards the employer brand image promoted through the culture of the organization."

According to Harquail (2006), "[t]he action implied by the label employee branding is meant literally because these programs are intended to impress brand attributes onto the work behavior of employees, who are then expected to infuse brand attributes throughout their work."

In general, a positive image of an organization represented to customers and other stakeholders by employees is called employee branding. The role of employees in building a brand is getting favored by organizations. Organizations attempt to influence the employees to project brand core value in a positive way to potential customers. To influence the employees toward brand, there is a need to connect employees' interest with organizational goals so that they can feel connected. Employees should feel like they are brand employees, the organization in which they are working is a brand, and it can fulfil their needs and wants. Employees should feel satisfied and proud

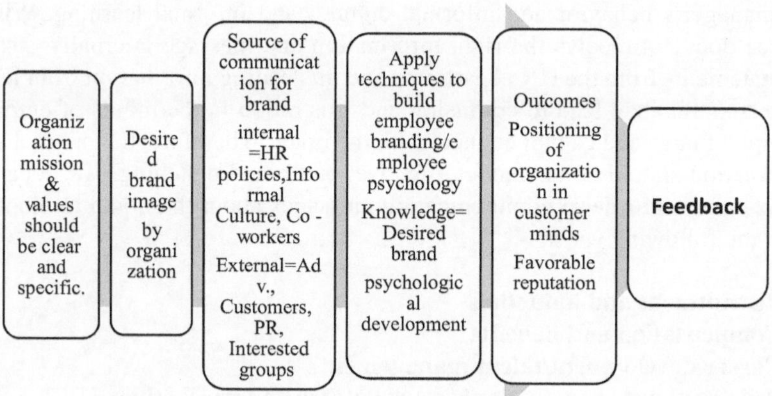

FIGURE 7.2 Model of process of employee branding.

(Source: Authors' own work.)

to be part of an organization. There are so many ways to make employees as brand employees.

Process of Employee Branding

There are different ways and techniques through which employee branding can be built and promoted, but there is one specific process to create branding through workforce. The below mentioned model is an inference through the model defined by Miles and Mangold's six-step process.

Figure 7.2 shows the process of employee branding that empowers the organization to solidify the image of their brand in the minds of customers and employees alike. The process clarifies the steps and ways to strengthen the desired image through employees, which is as follows:

1. Clear Organizational Goals and Desired Brand Image: This process says that firstly an organization should be clear about its mission, values, and desired image of the brand. What is the objective behind the brand and what would be its image in the market should be clear to the organization, and then only they will be able to educate employees about it.

2. Effective Communication: Organizations need to communicate employee branding through proper channels either internally or externally.

Internal Sources

Internally, employees can be communicated through the human resource (HR) management system – the compensation system which is a strong tool to match employees' goal with the organizational goal, training and education program, and organizational culture tell a lot about the brand and its image; co-workers' activities put a better influence on an employee and a new employee's behavior, leader/manager's behavior and informal culture, and informal learning. When an employee does not receive the right information and message internally – such as vague statements from the HR department and misleading information from leaders/coworkers – this may lead to confusion and frustration in the minds of employees about brand image and cannot connect with the brand. Therefore, the internal source of information should be authentic, effective, and brand matching. HR department plays an important role in communicating messages about the brand to employees through the following ways:

- **Recruitment and induction**
- **Compensation and benefits**
- **Career development/talent management**
- **Policies about organization's work culture and work ethics**
- **Employer value proposition**
- **Annual reports and meetings**

External Sources

Communication also helps to transmit information about the brand to existing employees, new talent, customers, and stakeholders. Organizations hire public relations agencies to build the brand image of the organizations as well as the products they are offering; advertisement is the common and an effective way to convince about a brand; customer feedback and word of mouth also represent another way. The external source of communication is a traditional approach that organizations usually follow to do branding of their product. This external source of information should match with the internal source of information for better consistency of the brand in the eyes of the employees.

Whether externally or internally, the source of information provided by an organization should be effective, consistent, and capable of convincing the customers as well as the employees about the brand.

3. Building Employee Branding: After communicating properly about a brand to employees, there is a need to connect the employees with the brand through psychological aspects. The organization should take initiatives to maintain a psychological connection with the employees through the following steps:

1. Informal learning programs where employees can have free discussion about the brand and their work with leaders and coworkers
2. Equal participation from each employee in decision-making.
3. Team work
4. Equality in all aspects (levels and their benefits/responsibilities)
5. Feedback from employees

Building brand employee is an important step of the process to connect employees mentally and emotionally with the brand so that they can connect the brand to customers further. Through connecting employees with the brand, organizations can create the desired brand image in the minds of employees.

4. Outcomes: When employees connect with the brand successfully, there is a need to check that it results in the desired output: building organizational positioning in the market effectively, increase in productivity, reduction in employee turnover, and attraction of new talent.

5. Feedback: Each and every process is followed by a feedback process to check whether process activities are properly implemented and the desired results are achieved. If it lacks in achieving any of the outcomes, then again follow up with the first step and check it properly. It is a continuous process.

Importance of Psychological Aspects in Employee Branding

Psychological aspects indicate mental relation between an employer and employee, a series of expectations of both from each other. Employees expect employers on the basis of information and message they receive from the beginning to retirement. If information and promises given by the employers in front of the employees are real, then a strong psychological relation can be developed between the employer and the employee. Strong psychological relation means trust, satisfaction, loyalty, efficiency, and positive words are communicated. On the other hand, if employees feel that there is no fairness, equality, and quality life, then it will bring negativity, less loyalty, trust, and negative words are communicated.

The psychological aspect plays a vital role in building the brand image in the minds of employees. For example, employees join a company for personal growth and after joining if that organization gives ample opportunity for their growth, it will develop a psychological relationship between the employer and the employee which will be long-lasting. When employees receive information about the organization and experience it, it makes an image of the organization. What image employees build in their minds is communicated to others (e.g., customers, new employees, and interested people). If they perceive a positive image from their experience, they reflect a positive image of the brand. If they perceive a negative image, they reflect a negative image from their behavior. Therefore, it is very important to internalize the image of the brand through employees. The internalization of brand helps the employees:

1. To understand the job of employees
2. To take a job-related decision
3. To build a positive behavior in employees toward the brand
4. To project a positive image in the minds of customers
5. To understand in a better way the customer's needs, and expectations
6. Know the organization's benchmarks for customer's services

Building psychological relation between the employer and the employees through internalization helps to satisfy the customer expectation and to raise the bar of an organization. Psychological relations can be developed through some measures and techniques (Rana et al., 2016).

Employee Banding and Brand Image

Employee brand image means the image perceived by employees about the organization's brand, that is, the degree to which they have internalized the brand image inside them. Employees adopt a brand only when they have a high-level sense of belongingness with the brand. When employees feel that this brand is mine, they present that brand to others as their child. It is only possible when they play an important role in the formation of that product. Every employee from each level or department can be the brand builder and projector if that employee is treated as an important part of it. The organization should take some strategic decision to make an image of the organization in the eyes of employees and some are trying to do that. In the next section, there are some of the techniques and practices discussed to create employee brand image.

Techniques for Employee Brand Image

There is no need for technical expertise to create employee brand image; only small practices can help the employer to make their employees as brand ambassador for their product, such as:

1. Teaching Employees: The step towards employee branding is to get them to know what is their brand, what is its purpose, and for what they are working. Training programs should be organized by the organization with the help of experts so that they can learn about the product and also how to communicate about it flawlessly in the market. It is a necessary and conventional technique to make employees part of the branding.

> Some of examples through which an organization teaches about the brand are role play, simulation training, presentation with videos and visuals effects, and buddy up approach.

2. Employees Branding Practice: After making an understanding of the brand to employees, there is a need to check their thoughtful ways of communicating it to customers. Organizations send employees in the market to project their brand to real customers, both existing and potential. This is an additional step where employees are taken outside of offices and that puts them in situations where they come face to face with the customers. Before the organization's sales team need to practice this, the organization prefers that every employee spends a few days with actual customers so that they can connect the brand identity with their identity.

> For example, the organization arranges workshops, seminars, and conferences where they assign a few employees to interact with customers. One of the best ways to make the employees understand the brand value in the market and customer's perception is by ghost shopping. **Ghost Shopping** is something where employees go to retailers, dealers, and stores as customers and do enquiry about the brand and value of the brand.

3. Employee Engagement: Explaining employees about the organization brand and putting it in the front of an interest group is the foundation of making employees part of the branding process. But to connect employees with a brand is not easy, for that employee's engagement is needed at each level or we can say employee's participation is required. Employees cannot be treated as a human-machine, only order them and they get work done. This is not the right approach; each employee should be treated as an important element of the organization. Equal opportunities should be provided to all employees in working, responsibilities, wages, benefits, and involvement in decision-making so that they can feel like part of that brand and treat the brand identity as their own identity and present it to the world as their own identity.

For example, practices like MBO, proper distribution of responsibilities and authority, job rotation, and transparency in action give encouragement for employee's participation. **Microsoft** adopted employee engagement as a strategy to build its brand image.

4. Happy Working Environment: If employees feel they are working for their own and at their place, it helps them to build a strong bonding with the brand. Employees not only work for remuneration, but also for a healthy and happy environment that provides them satisfaction. A satisfied employee always feels proud to announce his/her organization and brand for which he/she is employed. If employees are happy and satisfied with their work, an organization can convince customers more efficiently and effectively.

For example, informal culture/learning, less restrictions, owner of their own work, small-small appreciations, offsite meetings and parties are some practices which bring happiness to employees. **Zappos** works on the happiness of employees first which automatically converts it into customer happiness and creates the brand for them. **Next Jump** rewards employees to build the culture of the organization and ensures that all employees feel like a family at workplace.

5. **The Hiring of People:** At the time of recruitment, look at the candidate who can align with the organization's brand and also whether the candidate matches the company's values and can adjust to the environment and the current employees. When an organization hires employees, who match with the organization's desired brand image, it is easy for the employer to educate and train them further. Therefore, some organizations conduct informal interviews with candidates to know the real side of them and their social behavior.

For example, **AirBnB** does final selection on the basis of whether a candidate is aligned with their values and has the ability to deliver the brand to others. **Zappos** conducts one interview form to know if a candidate is able to create happiness at workplace.

6. Sense of Belongingness: Make employees feel that the workplace is theirs and they are the boss of their work, desk, decisions, and clients. So, this ownership makes them innovative, responsible, connected, and feel as a true representative of their work (brand/organization).

For example, Google shows that brand must be created inside; it follows the learning pattern of 70:20:10 which gives employees space for innovation.

7. Drive Communication: Whether new or existing, it is important to create an open communication forum. Organizations should communicate frequently and clearly about the mission, values, benefits, and desired brand image to all the workforce (lower, middle, and upper-middle employees). Always appreciate feedback of employees because they are in direct contact with customers and thus better understand what the desired brand is. Create such an environment where employees can share their experience without hesitation and feel that they have a stake in sharing and building the brand image.

"Value Employees' Feedback"

For example, more initiatives should be given towards informal social events so that free, casual talk and discussions can be appreciated. Google, Infosys, IBM, etc., follow informal learning practices.

8. Growth of Employees: As companies think about their growth, in the same way employees are always concerned about their growth. If companies show a lack in their concerns for employees' careers, then employees stop bothering for their companies' growth. It is a give and take relationship between an employer and an employee.

For example, from time to time, incentive programs, promotions, degree programs, online courses, etc., are offered for the employees, which are helpful for employees' future.

9. Reduction in Levels: Hierarchy and employment levels create discrimination among employees; therefore, it is better to divide them into teams where they can work at the same level as part of a team. In a team, an employee feels as a role-player and makes the best effort to complete the task efficiently. Teams give more opportunities to all the members to participate and work in their way.

For example, networking structure should form rather than hierarchal or line and staff structure.

10. Employees as a Customer: Organizations should treat their employees as their first customer so that they can contribute to improvising the brand. Whatever the new changes or product launches, make employees to be the tester for that so that they can give good advices or feedback, which can help in making necessary changes at the right time. In this, employees also feel special and as product inventors.

For example, distribution of samples to employees by an organization for approval before launching the product in the market.

These are the techniques through which an organization can answer employees' questions like:

1. What is this brand?
2. What are its attributes?
3. What the brand can do for me?
4. Whatever the brand promises that it fulfills or not?
5. Are my desires getting fulfilled through the brand?
6. Whether the brand matches with their identity or not?
7. Will the brand help to develop their future?

With these techniques employers also will be able to understand their employees and their expectations, and can build their psychological connection with them. Psychological integrity between the brand and employees helps in to build a better brand identity.

Some other small practices such as honesty towards employees and work, fairness and equality for all, harmony among employees, focus on learning of employees through informal means, and importance to employees' interests help in building employee branding.

Here are some of the examples of employee branding practices followed by big business houses:

- SATYAM: Every satyamite is a leader
- CTS: Celebration at work
- ACCENTURE: Best place for women to work
- LG: Best employee bonus
- MARUTI: Collective vacation scheme
- GE: Engineers with sense of humor

Without a right, trained, engaged, satisfied, and aligned, connected employee the organization cannot achieve its brand goal. According to D. K. Srivastava, vice president of HCL, "[a]n unsatisfied customer tells ten people about his/her experience while an unsatisfied employee tells a hundred." In employee branding, employees become vocal of the employer brand, simply put all attributes of the brand absolutely to interest people and they become a true brand ambassador of the employer. Employee branding does not require much effort to be put, but it is only to infuse brand attributes into the behavior, working style, communicative process, the environment, and most importantly into the minds of employees so that they can become brand propagator. An employee is the only source who can tell about the organization with emotions, feelings, and connectivity to the public.

Here are some of the few examples of organizations that built brand image among employees first and then through that market brand image. How employees rate their brand is the way they can convince others for it.

Microsoft, where 84 % of the employees feel proud to be part of it and 83% employees found that they work in a positive environment. Microsoft generates a score card where employees give rating to it: 1. On the basis of employee's retention, 2. On the basis of culture, and 3. On the basis of happiness. As per the employees in 2020 report card, Microsoft gets an A for the retention of employees, B for culture, and A+ for creating happiness among employees.

Google is called the organization for people. It is known for its work culture, perks and benefits, and giving value to creativity. Google provides a few things which make it a brand for their employees, such as meals at working time, relaxation time, and gives employees a chance to choose project by their own interest. All this helps Google to connect employee branding with the external branding, which projected it as an employer with the most talented workforce who can serve the world with best innovative technology and quality life with quality work.

Tata Steel is the first company in India which worked as a pioneer for employees' welfare schemes even before it was mandatory for industries and treat employees as a strategic partner in business.

BRANDING THROUGH WORKFORCE

Brandful Workforce-employees can be the most important component of your brand success.

(Julia Gometz, The Brandful Workforce)

Employee branding automatically boosts the employer brand. Employee branding is an authentic, innovative, and helpful tool for endorsing brands in many ways. Employees are the most effective influencers who can positively present the brand. Despite the branding, it also provides benefits to the whole organization.

1. Employee Identity: Employee branding gives identity to the employees as a brand creator and communicator. It helps employees to understand the core value of the brand for which they are working. They can easily communicate with others who are they and for what they are working. Proper education about brand and employee's role in it help to create employee identity.

2. Sense of Ownership: Employee branding develops a sense of ownership in employees, and they feel that the brand is their creation. Through providing a family environment and participation of employees, it can generate the feeling of ownership in employees.

3. Positive Behavior: It helps to shape the positive behavior of employees toward the brand. When employees feel connected to the brand and their organization, they feel positive about the brand for which they are working. Through proper training, attributes of a brand can be explained to the employees.

4. Brand Ambassadors: Employees can be the full-time brand ambassador of product through their work, communications, meetings with clients, presentation while solving queries of customers, and convincing new customers. It is only possible when they are well trained about the brand.

5. Brand Identity: When employees understand clearly about the brand, they can establish brand identity among customers. Employees project brand identity to others through their work, public behavior, and social media behavior.

6. Mutual Interest: While building employee branding, employers connect their interest with employees' interest, which helps in to develop a mutuality among themselves.

7. The Persona of Organizations: With the brand employees personality of organization can develop, that is, what their organization is all about. Through employee banding personality of the company, the company's culture, its management policies, the working style of the company, and the growth of a company can be created.

8. Retention of Employees: The process of employee branding gives satisfaction to employees, which cannot be achieved only by receiving a handsome salary. Satisfaction comes when they get recognition, find the importance of their own, and opportunities for growth. Through employee branding employers branding can be created which makes employees satisfied and retain them for a long time. It reduces the employees' turnover ratio.

9. Attract New Employees: If your current employee is happy with his/her brand, they tell about others for a vacant position in the organization. A good brand always attracts new skilled talent in the organization, when it is a propaganda by brand employees. Employee branding helps to fill a vacant position with the best talent.

10. Reduce Marketing Cost: When the organization believes in branding through its workforce, it does not need further investment in other ways of branding. The branded workforce express well about the brand to the customers and motivate them to purchase their product. Employee branding helps to build loyal customers and attract potential ones. It reduces the marketing and branding cost of the organization.

11. Reduce Hiring Cost: Less turnover ratio leads to less hiring and training programs in the organization. The brand image which is formed by employees helps to call the right applicant for the vacant post that reduces the costing involved in the recruitment process. If employees are happy with their brand, they refer to others too. Indirectly, employee branding reduces the cost of the new hiring process, training, and retention programs.

12. Innovative Approach: Branding through employees is a creative and innovative approach; it is different from traditional methods. It is innovative because it involves an employee who is having its own style to perform the things, which brings innovation in branding. An employee does branding of organization with customers in the following way:

1. Maintaining good relationship with customers (sometimes customers don't know about the organization well, they know employees who create the product and tell about it)
2. Communication skills

3. Working style
4. Ways to approach new customers
5. Through their behavior
6. By customizing the goods and services for customers
7. Through their social media account

13. Connect Customers: Employee branding helps to connect the actual brand. Through external branding, the customer gets only about the use of products, discounts, and features, but with employee branding, customers know the reason behind the brand, the core value of the brand. It connects customers with the actual brand.

14. Favorable Reputation: Employee branding develops a favorable reputation of the employer not only among customers but also among other stakeholders. Employees do not represent and contact customers only, but they also deal with suppliers, retailers, contractors, third parties, and mediators.

15. Positive Mouth Publicity: A branded employee speaks about their brand not in working hours but 24*7. Word of mouth is the biggest publicity; through employee branding, it can be created by organizations for customers.

> **For example, an employee of** Cisco mentioned about it that "I have complete autonomy to do what's best and if I face some problem need help, [sic] I have a strong support system from my team. I love working for Cisco and hope to continue for a very long time. I also love the diversity in the executive leadership, which is unique."

(www.greatplacetowork.com/best-workplaces/diversity/2019) Employee branding helps his employer to project themselves as a brand of their workforce in the minds of potential investors, recruiters, vendors, suppliers, retailers, and customers. Employee branding helps to create an internal brand representation that can talk about it, can live it, can handle it, can customize it, can give real feedback of it, and above all can represent it as an owner. An organization with an employee not only creates brand ambassadors for themselves but also is able to set up their brand in the market with this unique process, that is, employee branding.

LIMITATION OF BRANDING THROUGH WORKFORCE

As we discussed branding through employees and its importance, with that it is also important to put light on some drawbacks of it. Here are some limitations such as:

- Human behavior is unpredictable; it is difficult to guide their behavior for building a brand image in their minds.
- It is tough to harmonizing the diversity of the workforce, as they belong to different groups such as age, gender, religion, state, country, etc.
- It is not easy to synchronize workforce interest with an organization's interest because everyone has different interest.

- It is a time-consuming process.
- Employee branding requires an employee's involvement, which can blur the distinction between management level and employee level.

Every process has some pros and cons; the same employee branding process has some limitations which can be controlled through understanding the behavior of individuals and groups. However, dividing the whole process into small-small practices will help to save time and bring more effectiveness. Informal culture will help in to make employee involvement.

CONCLUSION

Employee branding is a win-win situation where both the employer and the employee get benefitted. Employers are able to get their brand image in the market in a more effective way; on the other hand, employees get recognition of their work in the form of calling them as a Brand Ambassador. The role of human resource is important in brand building because this is the only source among all which can talk, present, show, and mold the brand image of the organization. Employee branding can be a strategic tool that could have competitive advantage for organizations (Rana & Sharma, 2018). Through employee branding employers can get both loyal customers and motivated employees, which is the key for the long-term success of a business in a competitive market. Top management and HR practitioners should make note of this, that is, to promote their brand from each stakeholder's perspectives. Employee branding can be the new mantra for branding if proper processes and techniques are followed by the organizations. Before this, it was the belief that great brand tells the story to their prospect's talents of organizations, but nowadays it is found that real stories of the brand come from employees. Employees tell that what is this brand, why is this brand (Mission and Vision) and only why this brand to the customers. Most importantly, employee branding changes the behavior of employees, which is not easy to make. Employee branding will be the right branding method if proper processes, techniques, and remedies to overcome limitations followed by the organizations. Future saying will not be "brand your organization" but "brand your employees." Through the above discussion, further work can be done on working model of employee branding according to the real-time situation of workplace. This is the new concept and approach in marketing; therefore, more research work and discussion are needed in this direction.

REFERENCES

Edwards, M. R. (2005). Employer and employee branding: HR or PR. *Managing human resources: Personnel management in transition,* Willey Blackwell, 266–286.

Ghosh, D. K., & Kulshrestha, S. S. (2016). Employee branding–becoming a new mantra for employee engagement. *International Journal of Science and Research*, 5(6), 2081–2086.

Harquail, C. V. (2006). Employees as animate artifacts: employee branding by "wearing the brand". In Rafaeli, A. and Pratt M. (Eds.) *Artifacts and organizations: Beyond mere symbolism*, pp. 161–180. Mahwah, NJ: Lawrence Erlbaum Associates.

Harquail, C. V. (2007). Employee branding: Enterprising selves in the service of the brand. *Journal of Management*, 23(4), 925–942.

Kolachi, N. A. (2013). Competitive branding & development model: A qualitative case study of UAE approach to human capital. *Bulletin of Education and Research*, 35(1), 95–105.

Marchington, M., & Wilkinson, A. (2005). Direct participation and involvement. Managing human resources: personnel management in transition, 398-423.

Månsson, T., Törnqvist, N., & Erik, M. (2010). Employee branding at a pharmaceutical company (Thesis paper). Available at: www.diva-portal.org/smash/get/diva2:354127/FULLTEXT01.pdf (accessed November 2020).

Mangold, W. G., & Miles, S. J. (2007). The employee brand: Is yours an all-star? *Business Horizons*, 50(5), 423–433. doi:10.1016/j.bushor.2007.06.001

Miles, S. J., & Mangold, G. (2004). A conceptualization of the employee branding process. *Journal of Relationship Marketing*, 3(2–3), 65–87.

Patla, S., & Pandit, D. (2012). Internal branding in an Indian Bank: An initial exploration. *Vilakshan: The XIMB Journal of Management*, 9(1).

Pilotta, J. J., Schultz, D. E., Drenik, G., & Rist, P. (2004). Simultaneous media usage: A critical consumer orientation to media planning. *Journal of Consumer Behaviour: An International Research Review*, 3(3), 285–292.

Rana, G., & Sharma, R. (2019). Assessing impact of employer branding on job engagement: A study of banking sector. *Emerging Economy Studies*, 5(1), 7–21. https://doi.org/10.1177/2394901519825543.

Rana, G., Rastogi, R., & Garg, P. (2016). Work values and its impact on managerial effectiveness: A relationship in Indian context. *Vision*, 22, p. 300–311.

Rana, S., & Sharma, R. (2018). An overview of employer branding with special reference to Indian organizations. In: Sharma, N., Singh, V. K., & Pathak, S. (Eds.), *Management techniques for a diverse and cross-cultural workforce* (pp. 116–131). IGI Global USA. http://doi:10.4018/978-1-5225-4933-8.ch007.

Rosilawati, Y. (2014). Employee Branding Sebagai Strategi Komunikasi Organisasi Untuk Mengkomunikasikan Citra Merek (Brand-image). *Jurnal Ilmu Komunikasi,* 6(3).

Semnani, B. L., & Fard, R. S. (2014). Employee branding model based on individual and organizational values in the Iranian banking industry. *Asian Economic and Financial Review*, 4(12), 1726.

Sharma R., Singh S. P., & Rana G. (2019). Employer branding analytics and retention strategies for sustainable growth of organizations. In: Chahal, H., Jyoti, J., & Wirtz, J. (Eds.), *Understanding the Role of Business Analytics*. Singapore: Springer. https://doi.org/10.1007/978-981-13-1334-9_10.

Verghese A. K., (2015). Brand your employees, not your organization, ARK Group's Strategic Internal Communication 2015, Melbourne, Australia.

www.talentintelligence.com/blog/employee-branding-what-could-it-mean-for-your-organization

www.greatplacetowork.com/best-workplaces/100-best/2020

www.comparably.com/companies/microsoft/employer-brand

8 Impact of Knowledge Management on Employee Satisfaction in Nepalese Banking Sector

Sajeeb Kumar Shrestha

INTRODUCTION

Human resources run today's organizations, but large organizations are run by knowledge. So, knowledge-based perspectives in an organization give more attention to human creativity, skills, and competencies for the betterment of the organizations (Kianto, Vanhala, & Heilmann, 2018). These days, people are viewed as human capital, and their creativity and competences matter for the organization's long-term sustainability. So, knowledge management is viewed as a critical factor for business success in the competitive business world (Alyoubi, Hoque, Alharbi, Alyoubi, & Almazmomi, 2018). Human knowledge is considered a prime factor than financial and technical factors in the modern business world (Malhotra, 1998).

Knowledge Management is the identification and influence of accumulated knowledge in the organization to compete differently (Von Krogh, 1998). Knowledge management encompasses of knowledge acquisition, knowledge creation, knowledge sharing, knowledge storage, and knowledge transfer efficiently through management process, infrastructures, and system (Alavi & Leinder, 2001; Davenport, & Volpel, 2001; Lee & Choi, 2003; Rana, 2010; Sabherwal & Sabherwal, 2005). Nonaka and Takeuchi (1995) depict that knowledge management processes are interlinked processes. It is viewing as knowledge related outlook of an organization that elaborates mainly on individual people as human assets and people's expertise, knowledge, manner, and competences, and motivation for organizational upliftment (Rana & Goel, 2017; Schultz, 1961).

Knowledge management influences employee performances as well as organizational performances in the positive direction (Rasula, Vuksic, & Stemberger, 2012). Financial sectors are witnessed in undertaking knowledge management activities extensively for internal communication and effective utilization of knowledge related information (Chigada & Ngulube, 2015). Knowledge management dimensions are viewed as determinants for making competent employees, motivating them, and leading to employee satisfaction (Kianto et al., 2018; Alyoubi et al., 2018; Arif & Rahman, 2018; Ghanbari & Dastranj, 2017).

LITERATURE REVIEW

Knowledge is the data that should be processed in a meaningful way to convert information (Sabherwal & Fernandez, 2003; Soo, Devinney, Midgley & Deering, 2002). Balmisse, Meingan, and Passerine (2007) explored explicit knowledge and implicit knowledge.

Explicit knowledge represents words and numbers that are accessed easily (Balmisse et al., 2007). Implicit knowledge is insights and understandings that are difficult to share and difficult to access (Balmisse et al., 2007). Application of knowledge management results in increasing employees' performance and providing a competitive advantage in the organization (Bogner & Bansal, 2007; Rana, Rastogi, & Garg, 2016; Sabherwal & Sabherwal, 2005)

Knowledge acquisition means gathering all related information from external sources (Cohen & Levinthal, 1990). The organization has maintained a systematic network and communication channel to obtain information from a variety of sources. Organizations manage management information systems that are integrated with the employee and consumer comments, data analytics, market intelligence, and external partnerships to acquire the organization's information regularly.

Knowledge sharing means the exchange of personal and organizational knowledge and experiences. Frappaolo (2006) argues that organizations should manage the environment of knowledge between inter and intra individual, groups, and organizations. Knowledge sharing should be a voluntary social process to acquire, share, store, and retain (Harder, 2008). Senge (1997) depicts that sharing knowledge means caring to others.

Knowledge creation is developing new ideas and solutions to solve organizational problems and improve employee performance in turbulent times (Nonaka, Toyama, & Konno, 2000; Teece, Pisano, & Schuen, 1997). Commonly, members of an organization interact with each other to innovate knowledge. Scharmer (2001) depicts that organizations should facilitate unique knowledge development at all levels.

Knowledge storage is the keeping and arrangement of critical knowledge for current and future purposes and should be able to retrieve easily (Lawson, 2003). Primarily tacit knowledge should be coded or documented into explicit knowledge (Filius, de Jong, & Roelofs, 2000). Available technology should be suitable for employees and have the technical know-how to handle it.

Knowledge retention means attracting and retaining more knowledgeable and talented people in organizations. Organizations also face challenges of managing employee turnover and loss of expert knowledge (Kianto et al., 2018).

Employee satisfaction means an employee in an organization either likes or dislikes his job (Spector, 1985; Weiss & Cropanzano, 1996). Employees have different feelings towards his/her job when employees compare their expectation to real outcomes from the job (Cranny et al., 1992; Smith et al., 1969). It is an emotional feeling of an employee (Oshagbemi, 2000). Internal and external variables affect employee job satisfaction (Misener, Haddock, Gleation, & Ajamieh, 1996). Rana and Sharma (2019) argue that employees are the organization's internal markets; they could be nurtured as brands for job engagement.

Alyoubi et al. (2018) measured the impact of knowledge management on employee work performance as evidence from Saudi Arabia. The study's primary objective was to assess the effect of knowledge management processes as knowledge acquisition, knowledge sharing, knowledge creation, and knowledge retention with knowledge management approaches like social networks, codification, and personalization on job satisfaction. A causal model was proposed. Descriptive and explanatory research design was used. Survey research was done. Data were collected from employees from King Fahd National Library in Jeddah, Saudi Arabia. One hundred four samples were valid. Structured questionnaires were administered, and it is designed on five-point Likert scales, showing "1=Strongly Disagree," "2=Disagree,"

"3=Neutral," "4=Agree" and "5=Strongly Agree." Partial least square structural equation modeling was used to test the psychometric and econometric aspects of the causal model. It was found that knowledge sharing, knowledge retention, and knowledge codification/storage had a significant effect on employee satisfaction. In contrast, knowledge acquisition and knowledge creation had no significant effect on employee satisfaction. Alyoubi et al.'s (2018) implication is that alignment of knowledge management processes are associated with knowledge management approaches to job satisfaction.

Arif and Rahman (2018) measured on knowledge management and job satisfaction. This research aimed to analyze the relation of knowledge management on job satisfaction in the healthcare sector in Malaysia. It was concluded based on qualitative data that knowledge management and job satisfaction are linked up with each other.

Kianto et al. (2018) analyzed on impact of knowledge management on job satisfaction in the Finnish municipal organizations. Descriptive and causal research design was used in this study. A total of 824 samples were collected in this study. Structured questionnaires were used for data collected. Survey research was done. Partial least square structural equation modeling was run to validate the model and test the hypothesis. Correlation analysis was done to examine the relationship between exogenous and endogenous construct. Moderation analysis was done for checking multi group analysis for general employees, experts, middle level managers, and top management in terms of tenure, age, and unit. Exogenous constructs of the study were knowledge acquisition, knowledge sharing, knowledge creation knowledge codification, and knowledge retaining and endogenous construct was employee satisfaction. This research showed that knowledge management parameters (knowledge sharing, knowledge codification/storage, and knowledge retaining) were the way to influence employee satisfaction in most employees. No support was confirmed for knowledge acquisition and knowledge creation in employee satisfaction. The implication of Kianto et al. (2018) is the alignment of employee job satisfaction with knowledge management activities in the organizational settings.

Popa, Stefan, Cristina, and Cicea (2018) examined the influence of knowledge management practices on employee satisfaction in the Romanian healthcare system. The major objective of this study was to analyze the impact of knowledge management facets on employee satisfaction in healthcare sector. Descriptive and causal research design was used in this research. Convenient sampling method was used for choosing sample in this research. Structured questionnaires based on Likert scales were administered for data collection. Survey method was applied. Four hundred and fifty-nine samples were collected. Exploratory factor analysis was done to extract factors and SEM was run to test hypotheses. Exogenous constructs were knowledge acquisition, knowledge sharing, knowledge utilization, and endogenous construct was employee satisfaction. It was found knowledge acquisition and knowledge utilization had influenced on employee satisfaction. Knowledge sharing had no influence on employee satisfaction. This implication of Popa et al. (2018) is that testing the causal model of knowledge management and job satisfaction.

Ghanbari and Dastranj (2017) studied the knowledge management perspectives on personnel performance on Hormozgan Payame Noor University. Major objective of this research was to examine the impact of create knowledge, acquisition knowledge, capture knowledge, transmission knowledge, application knowledge, and organizational knowledge on personnel performance. Responses were collected from 54 personnel. Research method was descriptive and explanatory and survey research was done based on structured questionnaires. Cronbach's alpha, Pearson's correlation, one-way ANOVA, and *t*-test were used for checking the consistency of questionnaires, testing the relationship between constructs, and for testing hypotheses. It was found that knowledge creation, knowledge acquisition, knowledge registration, knowledge organizing, transfer knowledge, and application of knowledge had impact on personnel performance.

Lawal, Agboola, Aderibigbe, Owolabi, and Bakare (2014) examined the knowledge sharing among academic staff in Nigerian University of Agriculture as a survey. The aim of the study was to analyze the knowledge sharing activities of the academic staff of the Nigerian University of Agriculture. Descriptive research was done. Data were collected from 267 samples. Random sampling method was used. Survey research was done. Structured questionnaires were the main sources for collecting quantitative data. Descriptive statistics of percentage and frequency were used for analyzing and summarizing the data. It was found that academic staff share knowledge daily. Academic staff were doing collaborative research who were working from different geographical locations. Internet is the major tools for sharing knowledge to outside the World. Academic staff viewed knowledge sharing as a powerful tool for sharing information among academic staff. Some staff have also low awareness about knowledge sharing among academic staff in the university.

Tajali, Farahani, and Baharv and (2014) studied the relationship between knowledge management on employees' performance and organizational innovation. Descriptive and correlation research was done. Data were collected from 101 managers and employees of Lorestan Province Telecommunication Company using random classified sampling method. Structured questionnaires were used for data collection. Measurement instrument is based on a five-point Likert scale. It was found that knowledge management was significantly correlated with employees' performance and organizational innovation in Lorestan Province Telecommunication Company.

Thakur and Sinha (2013) studied on knowledge management in India perspective. The purpose of this study was to identify the awareness of business organizations in implementing knowledge management in Bhopal. Descriptive research was done in this research. Structured questionnaires were designed for the survey. It was found that business organizations were aware acquainted about knowledge management. Business organizations have implemented knowledge management strategy. Those, who have not implemented knowledge management strategy was shown interest to implement in their business context.

Zargar and Rezaee (2013) studied the effect of knowledge management on the performance rate of employees. The major objective of the study was to understand the knowledge management and its effect on organizational success and employee performance. Three hundred samples were collected randomly. A descriptive and causal research design was used. It was found that knowledge creation, knowledge participation, knowledge organizing, knowledge application, knowledge performance evaluation, and know performance were correlated with each other in a positive direction.

In this way, in the context of Nepal, some studies such as Adhikari (2008), Gautam (2012), Chaudhary (2012), Khanal (2016), Khanal and Poudel (2017), and Biswakarma (2018) had highlighted n knowledge management perspective. Adhikari (2008) argued that educational institutions must be the hub for knowledge creation, sharing, collaboration, storage, and retention. Gautam (2012) examined knowledge management initiatives by faculties and found the faculties are making efforts to be competitive by themselves than institutional support. Chaudhary (2012) analyzed knowledge strategy in public and private sector banks and found a knowledge management strategy to align with human resource strategy gives better performance. Khanal (2016) studied awareness of knowledge management in Nepalese financial institutions and found awareness level if medium and some challenges in handling knowledge issues. Khanal and Poudel (2017) confirmed that knowledge obtaining, knowledge organizing, and knowledge application significantly influence employee satisfaction. Biswakarma (2018) found a positive relationship between knowledge creation and acquisition of employee job performance. Some studies were done in Nepalese perspectives, but rigorous quantitative model building for testing the psychometric and econometric aspects of the causal model was not done. For this reason, this study is carried out to examine the effect of knowledge management on employee satisfaction in the Nepalese banking sector. The following research questions were raised for this research,

- Does knowledge acquisition influence employee satisfaction?
- Does knowledge share influence on employee satisfaction?
- Does knowledge creation influence on employee satisfaction?
- Does knowledge storage influence on employee satisfaction?
- Does employee retention influence on employee satisfaction?

This research examines the effect of knowledge management facets (knowledge acquisition, knowledge sharing, knowledge creation, knowledge storage, and knowledge retention) on employee satisfaction in the Nepalese banking sector.

To show the relationship between constructs, the research framework is designed. Research framework shows the exogenous and endogenous constructs in the study. Figure 8.1 shows the research framework.

In this study, the following hypothesis has been set.

H1: Knowledge acquisition has a significant effect on employee satisfaction.
H2: Knowledge sharing has a significant effect on employee satisfaction.
H3: Knowledge creation has a significant effect on employee satisfaction.
H4: Knowledge storage has a significant effect on employee satisfaction.
H5: Knowledge retention has a significant effect on employee satisfaction.
H6: Gender moderates knowledge management dimensions to employee satisfaction

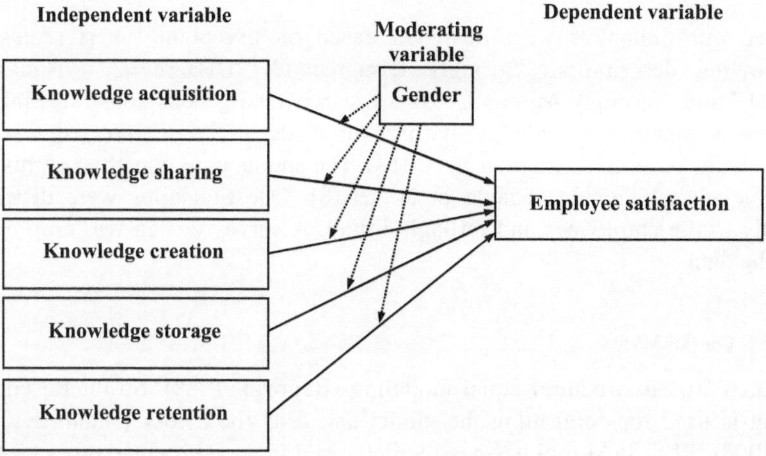

FIGURE 8.1 Research framework.

METHODOLOGY

Research Design

The objective of this study was to measure the impact of knowledge management on employee satisfaction. Malhotra and Birks (2006) suggest that descriptive and causal research design is suitable for achieving objectives mostly for research. Specific steps are carried to follow the research design. Descriptive research follows a formal study to review the constructs of interest (Cooper & Schindler, 2009) in detail (Malhotra & Birks, 2006; Zikmund, 2000). A causal research design is used to test the causal model.

Population and Sample

Employees working in Nepalese banks and financial institutions are the population of the study. The sample frame is the target population or subset of the population, and it is considered in organizations that reside in Kathmandu City. The study sample comprised 257 hundred employees who gave responses in the survey, and it is enough (Eldred, 1987; Kline, 2011). Two hundred samples are large, and it must be collected for using structural equation modeling and providing for better results (Eldred, 1987; Kline, 2011). Saunders, Lewis, and Thornhill (2003) argue that the sample size depends on research questions, objectives, and methodology.

Non-random sampling method was used for selecting the sample (Bryman & Bell, 2011). Judgmental sampling method was used. The employee working for more than six months in the banks and financial institutions is considered valid for this research.

Nature and Sources of Data

The nature of data was primary in this research. Employees working in the banks were approached personally or by e-mail.

Instruments and Survey

Structured questionnaires were designed based on five-point Likert scales with the following descriptors: "Strongly Disagree=1," "Disagree=2," "Neutral=3," "Agree=4," and "Strongly Agree=5." This research's exogenous constructs included knowledge acquisition, knowledge sharing, knowledge creation, knowledge storage, and knowledge retention (Kianto et al., 2018). The endogenous construct of this study was employee satisfaction (Kianto et al., 2018). Questionnaires were distributed physically to the employees and through e-mail. A survey design was employed to collect the data.

Methods of Analysis

This research used structural equal modeling (Bagozzi, 1988). Structural equation modeling is used for confirming the model and tests the theory (Babin, Hair, and Boles, 2008). PLS-SEM tool has been widely used in social science research to test the theory, and it has a typical application to predict the model (Henseler, Ringle, & Sinkovics, 2009; Hulland, 1999). Partial least square structural equation modeling (PLS-SEM) is a useful tool for handling complex data and also useful for small to medium sample size (Vinzi, Chin, Henseler, & Wang, 2010). Hayes (2020) Process Macros SPSS Version 3.5 was used for testing moderation analysis.

Demographic Profile of the Respondents

There were 51 % of male respondents and 45 % of female respondents. About 76% of the respondents were between 20 and 30 years of age and 20% were 30–40 years of age. The majority (70%) of the respondents were working in their organizations below 5 years and 24% were working for 5–10 years. Further, 65 % of employees worked at the assistant level, and 81 % of employees were working on officer level in their organizations.

RESULTS AND DISCUSSION

The partial least square analysis was done through Smart PLS 3software to test the proposed research model. SEM follows two-stage analysis procedures. First, the measurement model was tested for the suitability of the model's validity and reliability, and second, it tests the structural model (Hair, Hult, Ringle, & Sarstedt, 2013). The bootstrapping method was done through 5000 resampling to analyze the path coefficients of the constructs and their loadings (Hair et al., 2013).

Testing the Measure Model

Checking the Measurement Model Means

Earlier, the measurement model was checked for convergent validity. Composite reliability (CR), average variance explained (AVE), and factor loadings are the tools for testing the model's validity. Table 8.1 highlights all the outer loadings of indicator

TABLE 8.1
Validity and Reliability of the Model

Constructs	Items	Loadings	CR	AVE
Knowledge Acquisition	KA1	0.626	0.784	0.555
	KA2	0.665		
	KA3	0.911		
	KA4	0.780		
Knowledge Sharing	KS1	0.768	0.844	0.572
	KS2	0.823		
	KS3	0.728		
	KS4	0.710		
Knowledge Creation	KC3	0.658	0.862	0.612
	KC4	0.840		
	KC5	0.837		
Knowledge Storage	KST3	0.714	0.818	0.602
	KST4	0.865		
	KST5	0.739		
Knowledge Retention	KR3	0.756	0.853	0.660
	KR4	0.860		
	KR5	0.818		
Employee Satisfaction	ES1	0.721	0.882	0.652
	ES3	0.767		
	ES4	0.888		
	ES5	0.844		

(Source: Authors' own work.)

items are above 0.7 (Chin, 2010). Construct reliability explains the latent constructs if the threshold value is higher than 0.7 benchmark (Hair et al., 2013). AVE of all the constructs is reported more than 50% or 0.5 (Hair et al., 2013).

Table 8.1 presents that CR is higher than the 0.7 value for latent constructs, and it is greater than AVE. In conclusion, convergent validity is obtained.

In the second stage, the measurement model was tested for discriminant validity that tests the correlations between measures of interest and constructs' measures. Table 8.1 records that AVE of all the constructs is greater than 0.5 threshold value (Fornell & Larcker, 1981). It can be concluded that discriminant validity is obtained.

Further examining discriminant validity for the reflective measurement model, it was measured cross-loadings of the indicators. External loadings of constructs should be greater than loadings in row and column (Hair, Sarstedt, Hopkins, & Kuppelwieser, 2014). Table 8.2 shows the cross-loading of indicators of the related constructs. So, discriminant validity is further proved.

Heterotrait–Mnotrait ratio of correlations table is a useful benchmark for measuring discriminant validity in modern days (Henseler, Ringle, & Sinkovics, 2009).

Table 8.3 shows the square root of each construct's AVE in the diagonal line is greater than all the constructs' correlations in each row and each column. It is shown

TABLE 8.2
Cross Loadings of the Latent Constructs

Items	Employee Satisfaction	Knowledge Acquisition	Knowledge Creation	Knowledge Retention	Knowledge Sharing	Knowledge Storage
ES1	**0.721**	0.129	0.243	0.292	0.171	0.314
ES3	**0.767**	0.089	0.278	0.344	0.212	0.256
ES4	**0.888**	0.197	0.351	0.346	0.350	0.362
ES5	**0.844**	0.103	0.311	0.315	0.357	0.322
KA1	0.078	**0.626**	0.088	0.119	0.077	0.249
KA2	0.053	**0.665**	0.107	0.028	0.227	0.263
KA4	0.177	**0.911**	0.126	0.159	0.163	0.513
KC1	0.279	0.130	**0.780**	0.099	0.393	0.245
KC3	0.253	0.015	**0.658**	0.125	0.298	0.197
KC4	0.319	0.151	**0.840**	0.149	0.229	0.373
KC5	0.301	0.133	**0.837**	0.107	0.176	0.320
KR3	0.294	0.081	0.072	**0.756**	0.013	0.213
KR4	0.328	0.156	0.148	**0.860**	0.161	0.175
KR5	0.350	0.145	0.148	**0.818**	0.142	0.192
KS1	0.260	0.209	0.272	0.019	**0.768**	0.260
KS2	0.307	0.196	0.386	0.129	**0.823**	0.177
KS3	0.222	0.067	0.232	0.099	**0.728**	0.106
KS4	0.255	0.089	0.128	0.158	**0.710**	0.251
KST3	0.221	0.326	0.254	0.168	0.152	**0.714**
KST4	0.386	0.407	0.338	0.224	0.280	**0.865**
KST5	0.267	0.448	0.254	0.147	0.149	**0.739**

(Source: Authors' own work.)

TABLE 8.3
HTMT

	Employee Satisfaction	Knowledge Acquisition	Knowledge Creation	Knowledge Retention	Knowledge Sharing	Knowledge Storage
Employee Satisfaction	**0.808**					
Knowledge Acquisition	0.164	**0.745**				
Knowledge Creation	0.370	0.142	**0.782**			
Knowledge retention	0.400	0.159	0.154	**0.812**		
Knowledge Sharing	0.347	0.191	0.344	0.134	**0.759**	
Knowledge Storage	0.391	0.506	0.369	0.236	0.264	**0.776**

(Source: Authors' own work.)

in bold letters. All values are below the HTMT 0.9 criterions (Henseler et al., 2009). It can be concluded that there exists discriminant validity for the proposed model

Structural Model

The structural model is the proposed model for testing purposes. The hypothesis is testing, and the conclusion is made. Here, coefficient of determination (R2), beta of the constructs, t-value, and its significance are achieved through the bootstrapping algorithm of Variance Based Structural Equation Modeling to 5000 resampling. The structural model is expressed in Figure 8.2.

The coefficient of determination of the model is 0.335 or 33.5 %. This model can predict 33.5%.

From Table 8.4, it can be seen that knowledge acquisition did not significantly influence on employee satisfaction (β=-0.066, p> 0.05). Knowledge sharing influenced significantly on employee satisfaction (β=0.195, p<0.05). Knowledge

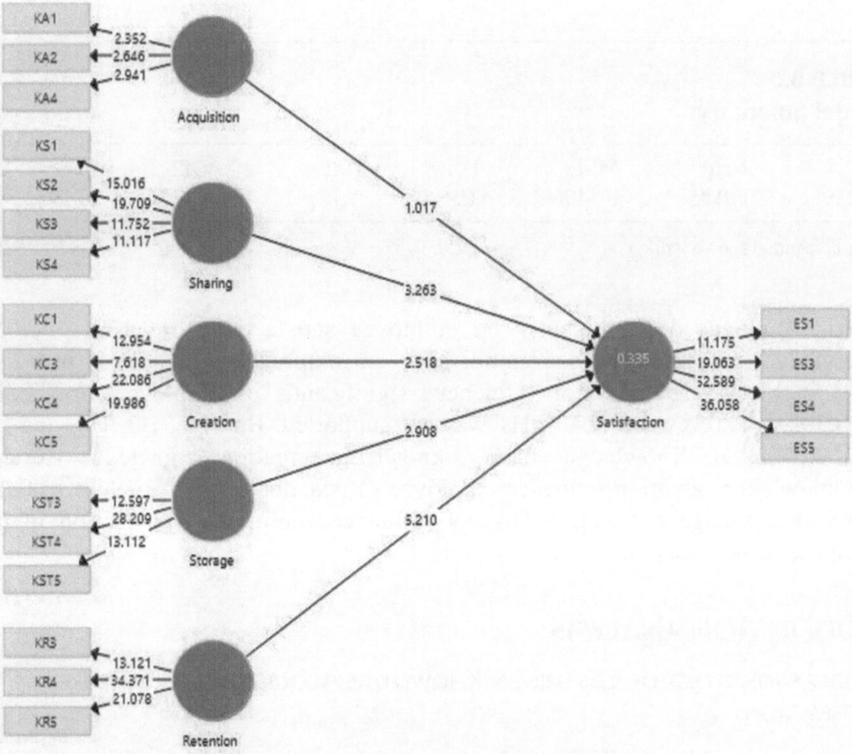

FIGURE 8.2 Structural model.

Note: Output of Smart PLS 3, where Acquisition = Knowledge Acquisition, Sharing = Knowledge Sharing, Creation = Knowledge Creation, Storage = Knowledge Storage, Retention = Knowledge Storage and Satisfaction = Employee Satisfaction.

TABLE 8.4
Hypothesis Testing

Hypothesis	Path Coefficients	T Statistics	p-value	Decision
H1: Knowledge Acquisition -> Employee Satisfaction	-0.066	1.017	0.309	Not Supported
H2: Knowledge Sharing -> Employee Satisfaction	0.195	3.263	0.001	Supported
H3: Knowledge Creation -> Employee Satisfaction	0.179	2.518	0.012	Supported
H4: Knowledge Storage -> Employee Satisfaction	0.235	2.908	0.004	Supported
H5: Knowledge Retention -> Employee Satisfaction	0.302	5.210	0.000	Supported

(Source: Author's own work.)

TABLE 8.5
Model Summary

R	R-sq	MSE	F	df1	Df2	p
0.1359	0.0185	0.4406	1.5861	3	253	**0.1932**

(Source: Authors' own work.)

creation influenced significantly on employee satisfaction (β=0.179, $p<0.05$). Knowledge storage influenced significantly on employee satisfaction (β=0.235, $p<0.05$). Knowledge retention influenced significantly on employee satisfaction (β=0.302, $p<0.05$). As a result, H1 was not supported. But, H2, H3, H4, and H5 were supported. Knowledge sharing, knowledge creation, knowledge storage, and knowledge retention influence employee satisfaction in the Nepalese banking sector. Knowledge acquisition has not influenced employee satisfaction in the Nepalese banking sector.

MODERATION ANALYSIS

Moderation effect of Gender on Knowledge Acquisition to Employee Satisfaction

Hayes' Process Macros Tool (2020) was used for checking the moderation effect of gender on knowledge acquisition to employee satisfaction. Here, Y=Employee Satisfaction; X=Knowledge Acquisition; and W=Gender.

Outcome Variable=Employee Satisfaction.

TABLE 8.6
Model of Interaction

	coeff	se	t	p	LLCI	ULCI
constant	3.4892	0.7816	4.4642	0.0000	1.9499	5.0284
Knowledge Acquisition	0.0656	0.2004	0.3275	0.7435	-0.3290	0.4602
Gender	-0.1685	0.5334	-0.3160	0.7523	-1.2191	0.8820
Interaction (Knowledge Acquisition × Gender)	0.0521	0.1360	0.3831	**0.7019**	**-0.2157**	**0.3200**

(Source: Authors' own work.)

TABLE 8.7
Model Summary

R	R-sq	MSE	F	df1	Df2	p
0.348	0.1215	0.3943	11.6680	3	253	**0.000**

(Source: Author Own)

The model summary was explained in Table 8.5.

Table 8.5 shows that the model is predicted by 1% only, and multiple correlation is 0.1359.

Table 8.6 shows the interaction effect of knowledge acquisition and gender on employee satisfaction; p-value of interaction is 0.7019, so it is not significant having zero value between lower and upper limits (LLCI=-0.2157 to ULCI=0.32000). Gender has no moderation effect on knowledge acquisition to employee satisfaction.

Moderation effect of Gender on Knowledge Sharing to Employee Satisfaction

Hayes' Process Macros Tool (2020) was used to test the moderation effect of gender on knowledge sharing on employee satisfaction. Here, Y=Employee Satisfaction, X=Knowledge Sharing, and W=Gender.

Outcome Variable=Employee Satisfaction.

The model summary was shown in Table 8.7.

Table 8.7 shows the model is estimated by 12.15%, and multiple correlation is 0.348.

Table 8.8 highlights the interaction effect of knowledge sharing and gender on employee satisfaction; p-value of interaction is 0.2480, which does not significantly have zero value between the lower and upper limit (LLCI=-0.3492 to ULCI=0.0906). Gender has no moderation influence on knowledge sharing to employee satisfaction.

TABLE 8.8
Model of Interaction

	coeff	se	t	p	LLCI	ULCI
constant	1.8780	0.6277	2.9918	0.0030	0.6418	3.1142
Knowledge Sharing	0.5019	0.1679	2.9895	0.0031	0.1713	0.8326
Gender	0.5240	0.4165	1.2581	0.2095	-0.2962	1.3442
Interaction (Knowledge Sharing × Gender)	-0.1293	0.1117	-1.1579	**0.2480**	**-0.3492**	**0.0906**

(Source: Authors' own work.)

TABLE 8.9
Model Summary

R	R-sq	MSE	F	df1	Df2	p
0.3686	0.1359	0.3879	13.2602	3	253	**0.000**

(Source: Authors' own work.)

Moderation effect of Gender on Knowledge Creation to Employee Satisfaction

Hayes' Process Mcros Tool' (2020) was used to measure the moderation effect of gender on knowledge creation on employee satisfaction. Here, Y=Employee Satisfaction, X=Knowledge Creation and W=Gender

Outcome Variable=Employee Satisfaction.

The model summary was highlighted in Table 8.9.

Table 8.9 shows the model is predicted by 13.59%, and multiple correlation is 0.3686.

Table 8.10 shows the interaction effect of knowledge creation and gender to employee satisfaction; p-value of interaction is 0.9089 (not significant) and having zero value between lower and upper limit (LLCI=-0.2398 to ULCI=0.2694). So, gender has no moderation influence on knowledge sharing to employee satisfaction.

Moderation effect of Gender on Knowledge Storage to Employee Satisfaction

Hayes' Process Macros Tool' (2020) was used to check the moderation effect of gender on knowledge storage on employee satisfaction. Here, Y=Employee Satisfaction, X=Knowledge Storage, and W=Gender.

Outcome Variable=Employee Satisfaction.

The model summary is explained in Table 8.11.

TABLE 8.10
Model of Interaction

	coeff	se	t	p	LLCI	ULCI
constant	2.1922	0.8015	2.7353	0.0067	0.6139	3.7706
Knowledge Creation	0.3808	0.1956	1.9466	0.0527	-0.0045	0.7660
Gender	-0.0222	0.5301	-0.0419	0.9666	-1.0662	1.0218
Interaction (Knowledge Creation × Gender)	0.0148	0.1293	0.1145	**0.9089**	**-0.2398**	**0.2694**

(Source: Authors' own work.)

TABLE 8.11
Model Summary

R	R-sq	MSE	F	df1	Df2	p
0.3655	0.1336	0.3889	13.0018		235	**0.000**

(Source: Author Own)

TABLE 8.12
Model of Interaction

	coeff	Se	t	p	LLCI	ULCI
constant	2.4212	0.8001	3.0262	0.0027	0.8455	3.9968
Knowledge Storage	0.3539	0.2141	1.6532	0.0995	-0.0677	0.7754
Gender	-0.1192	0.5056	-0.2358	0.8138	-1.1149	0.8765
Interaction (Knowledge Storage × Gender)	0.0438	0.1351	-0.3238	**0.7464**	**-0.2224**	**0.3099**

(Source: Authors' own work.)

Table 8.11 shows that the model is regressed by 13.36% and multiple correlation is 0.3655.

Table 8.12 expresses the interaction effect of knowledge storage and gender on employee satisfaction; p-value of interaction is 0.7464, which is not significant and having zero value between lower and upper limit (LLCI=-0.2224 and ULCI=0.3099). So, gender does not moderate knowledge storage to employee satisfaction.

Moderation Effect of Gender on Knowledge Retention to Employee Satisfaction

Hayes Process Macros Tool (2020) was applied to examine the moderation effect of gender on knowledge retention on employee satisfaction. Here, Y=Employee Satisfaction, X=Knowledge Retention, and W=Gender

TABLE 8.13
Model Summary

R	R-sq	MSE	F	df1	Df2	p
0.4011	0.1609	0.3767	16.1682	3	253	**0.000**

(Source: Authors' own work.)

TABLE 8.14
Model of Interaction

	coeff	se	t	p	LLCI	ULCI
constant	2.8440	0.4866	5.8453	0.0000	1.8858	3.8022
Knowledge Retention	0.3118	0.1481	2.1061	0.0362	0.0202	0.6034
Gender	-0.1060	0.3355	-0.3160	0.7522	-0.7667	0.5547
Interaction (Knowledge Retention × Gender)	0.0196	0.0998	0.1963	**0.8445**	**-0.1770**	**0.2161**

(Source: Authors' own work.)

Outcome Variable=Employee Satisfaction.

The model summary is shown in Table 8.13.

Table 8.13 shows the model is explained by 16.09%, and multiple correlation is 0.4011.

Table 8.14 highlights the interaction effect of knowledge storage and gender on employee satisfaction; p-value of interaction is 0.8445, which is not supported and having zero value between the lower and upper limit (LLCI =-0.1770 and ULCI=0.2161). In conclusion, gender does not moderate knowledge retention to employee satisfaction.

DISCUSSION

Knowledge management dimension on employee satisfaction is the interest of this study. The study is designed to measure the impact of knowledge management dimensions like knowledge acquisition, knowledge sharing, knowledge creation, knowledge storage, and knowledge retention on employee satisfaction in the Nepalese banking sector. Besides, the role of gender in moderating the relationship is also investigated.

Among the proposed five hypotheses, four were accepted. Knowledge sharing, knowledge creation, knowledge storage, and knowledge retention showed a significant positive influence on employee satisfaction (Alavi & Leidner, 2001; Davenport & Vopel, 2001; Kianto et al., 2018; Lee & Choi, 2003; Rana, 2010; Rana & Sharma, 2019; Sabherwal & Sabherwal, 2005). Knowledge management practices are critical tools for increasing employee job performance or employee satisfaction (Bogner & Bansal, 2007; Rana, 2010; Rana et al., 2016; Sabherwal & Sabherwal, 2005). There is room for nurturing employee as a brand and strengthening internal branding (Rana & Goel, 2017; Rana & Sharma, 2019).

Knowledge sharing has an impact on employee satisfaction. Knowledge sharing is strengthened in the organization through communication media, internet and e-mails, training, job rotation, and self-interaction with other employees, and use of technology makes smarter to share their knowledge (Alyoubi et al., 2018; Kianto et al., 2018; Popa et al., 2018; Rana & Sharma; 2019).

Knowledge creation has an impact on employee satisfaction. Knowledge creation is facilitated through on-the-job training; teamwork, open conversation, and expert consultation that makes employees independently handle their jobs (Alyoubi et al., 2018; Kianto et al., 2018; Popa et al., 2018; Rana & Goel, 2017).

Knowledge storage influences employee satisfaction. Organizations can store knowledge in the form of videos, documents, and the establishment of management information systems (Alyoubi et al., 2018; Darroch, 2005; Kianto et al., 2018; Popa et al., 2018; Rana, 2010).

Knowledge retention influences employee satisfaction. Knowledge is retained through nurturing employees and mentoring and coaching, attractive pay for smart employees, timely and fair promotion schemes, and other facilities that retain employees in the organizations for the long run (Alyoubi et al., 2018; Darroch, 2005; Lawal et al., 2014; Kianto et al., 2018; Rana & Goel, 2017; Zargar & Rezaee, 2013).

Knowledge acquisition does not create a change in employee satisfaction. Knowledge acquisition practices are poorly performed in the Nepalese banking sector, probably. This is in line with the findings of Kianto et al. (2018) but in contrast to Biswakarma (2018). This is due to the work context of the Nepalese banking sector that does not demand knowledge acquisition. The bankers do not promote knowledge acquisition activities like acquiring new knowledge, employee collaboration, and asking employees who are leaving the bank to teach others to acquire the knowledge (Alyoubi et al., 2018; Kianto et al., 2018).

Gender does not affect knowledge management dimensions to employee satisfaction. Male and female independently share, create, store, and retain knowledge.

This research is limited to primary cross-sectional data. Nepalese banking and financial institutions in Kathmandu City are evident for this research.

CONCLUSION

This research formulated five econometric causal relationships about knowledge management on employee satisfaction in Nepalese banking sector undertakings. This study aimed to measure the impact of knowledge management dimensions, causing employee satisfaction. Reliability and validity were tested to confirm the psychometric aspect of the constructs and its loadings. Findings suggest knowledge sharing, knowledge creation, knowledge satisfaction, and knowledge retention influence on employee satisfaction. This means that knowledge management practices make employees satisfied with their jobs in the Nepalese banking sector.

This research is consistent with Alyoubi et al. (2018), Kianto et al. (2018), Rana and Goel (2017) and Rana et al. (2016) in that knowledge management dimensions lead to employee satisfaction. This research contrasts with Biswakarma (2018) and Popa et al. (2018) in that knowledge acquisition does not support employee satisfaction. The Nepalese banking sector in these days is being competitive, and it needs professionals to perform the task and manage the surrounding organizational environment. From recruitment to retention is the key to the banking sector, so knowledge management initiatives should be taken proactively in the banking sector for successful planning and employee satisfaction.

IMPLICATIONS

Following managerial, research, and future research can be forwarded to apply the research findings and coming to the research context.

MANAGERIAL IMPLICATIONS

This research proved that knowledge management factors affect employee satisfaction. So, managers in the banking sector and other corporate sectors should implement knowledge management strategies for recruitment to retention to employees' career planning. Bank managers should apply knowledge management dimensions like knowledge acquisition, knowledge sharing, knowledge creation, knowledge storage, and knowledge retention in aligning with people, process, and technology basis. Talent hunt should be practiced that experts in the related field and professionals got an opportunity at the managerial level for formulating, developing, and implementing knowledge management facets efficiently.

RESEARCH IMPLICATIONS

Variance Based SEM (PLS-SEM) was used in this research. The researcher applied two-stage analysis procedures for checking the measurement model and testing the structural model. A psychometric test was done in the measurement model, and econometric testing was done in the structural model. Validation of this research model can be remedial for measuring knowledge management issues on employee performance in the different research tools contexts. Hayes (2020) process macros were also used to form moderation analysis of gender to knowledge management dimensions to employee satisfaction.

FUTURE RESEARCH IMPLICATIONS

The researcher has validated the model's psychometric and econometric aspects of reporting knowledge management and employee satisfaction in the Nepalese banking sector. As a result, this model can be used in other corporate sectors too. Gender is incorporated in this research model, whether it is instrumental for the banking sector or not. Besides gender, other variables like age, experience, and position can also be used to test the moderation effect on knowledge management and employee management context. Likewise, knowledge and employee performance should be extended to employee commitment, organizational performance, and other issues.

REFERENCES

Adhikari, D.R. (2008). *Knowledge management in academic institutions*. Kathmandu, Nepal: Campus Chief Seminar, Tribhuvan University.

Alavi, M., & Leidner, D.E. (2001). Review: Knowledge management and knowledge management systems: Conceptual foundations and research issues. *Management Information System Quarterly, 25*(1), 107–136.

Alyoubi, B., Hoque, Md. R., Alharbi, I., Alyoubi, A., & Almazmomi, N. (2018). Impact of knowledge management on employee work performance: Evidence from Saudi Arabia. *The International Technology Management Review, 7*(1), 13–24.

Arif, F.N.T., & Rahman, S.A. (2018). Knowledge management and job satisfaction. *International Journal of Academic Research in Business & Social Sciences, 8*(9), 266–274.

Babin, B.J., Hair, J.F., & Boles, J.S. (2008). Publishing research in marketing journals using structural equation modeling. *Journal of Marketing Theory & Practice, 16*(4), 279–285.

Bagozzi, R.P., & Yi, Y. (1988). On the evaluation of structural equation models. *Journal of the Academy of Marketing Science,* 16(1), 74–94.

Balmisse, G., Meingan, D., & Passerine, K. (2007). Technology trends in knowledge management tools. *International Journal of Management, 3*(2), 118–131.

Biswakarma, G. (2018). Knowledge management and employee job performance in Nepalese banking sector. *International Journal of Research in Business Studies and Management, 5*(3), 15–23. Retrieved from www.ijrbsm.org/papers/v5-i3/3.pdf

Bogner, W.C., & Bansal, P. (2007). Knowledge management as the basis of sustained high performance. *Journal of Management Studies, 44*(1), 165–188.

Bryman, A., & Bell, E. (2011). *Business research methods* (3rd ed.). New Delhi: Oxford University Press.

Chaudhary, M.K. (2012). *Practice of knowledge management strategy by banking industry of Nepal*. International Conference on Management, Humanity and Economics. ICMHE 2012. Phuket, Thailand. Retrieved from https://pdfs.semanticscholar.org/d648/c92cf659c1d3f46e47f201d920c82d4d21dc.pdf

Chigada, J., & Ngulube, P. (2015). Knowledge management practices at selected banks in South Africa: Original research. *South African Journal of Information Management, 17*(1), 1–10.

Chin, W.W. (2010). How to write up and report analyses. In: V. Esposito Vinzi, W.W. Chin, J. Henseler, & H. Wang (Eds.), *Handbook of partial least squares: Concepts, methods and applications in marketing and related fields* (pp. 655–690). Berlin: Springer Handbooks of Computational Statistics Series.

Cohen, W.M., & Levinthal, D.A. (1990). Absorptive capacity: A new perspective on learning and innovation. *Administrative Science Quarterly, 31*(1), 128–152

Cooper, D., & Schindler, P.S. (2009). *Business research methods* (10th ed.). New York: McGraw-Hill.

Cranny, C.J., Smith, P.C., & Stone, E.F. (1992). *Job satisfaction: How people feel about their jobs and how it affects their performance*. New York: Lexington Books.

Darroch, J. (2005). Knowledge management, innovation and firm performance. *Journal of Knowledge Management, 9*(3), 101–115.

Davenport, T.H., & Volpel, S.C. (2001). The rise of knowledge towards attention management. *Journal of Knowledge Management, 5*(3), 212–221.

Eldred, G. (1987). *Real estate: Analysis and strategy*. New York: Harper & Row.

Filius, R., deJong, J.A., & Roelofs, E.C. (2000). Knowledge management in the HRD office: A comparison of three cases. *Journal of Workplace Learning, 12*(7), 286–295. Doi: http://dx.doi.org/10.1108/13665620010353360

Fornell, C., & Larcker, D.F. (1981). Evaluating structural equation models with unobservable variables and measurement error. *Journal of Marketing Research, 18*(1), 39–50.

Frappaolo, C. (2006). *Knowledge management*. West Sussex: Capstone Publishing.

Gautam, D.K. (2012). Knowledge management initiatives by faculties of Tribhuvan University of Nepal. *International Journal of Business Performance Management, 12*(2), 158–166.

Ghanbari, S., & Dastranj, M. (2017). The effect of knowledge management on the performance of personnel of Hormozgan Payme Noor University. *Journal of Socialomics, 6*(4), 1–4.

Hair, J.F., Hult, G.T.M., Ringle, C., & Sarstedt, M. (2013). *A primer on partial least squares structural equation modelling (PLS-SEM)*. Los Angeles, CA: Sage Publications

Hair, J.F., Sarstedt, M., Hopkins, L., & Kuppelwieser, G.V. (2014). Partial least squares structural equation modeling (PLS-SEM): An emerging tool in business research. *European Business Review, 26*(2), 106–121.

Harder, M. (2008). *How do rewards and management styles influence the motivation to share knowledge?* In SMG Working Paper No. 6/2008, Available at SSRN: https://ssrn.com/abstract=1098881 or http://dx.doi.org/10.2139/ssrn.1098881

Hayes, A.F. (2020). *The process macro for SPSS, SAS, and R*. Retrieved from www.processmacro.org/index.htm.

Henseler, J., Ringle, C.M., & Sinkovics, R.R. (2009). The use of partial least squares path modeling in international marketing. *Advances in International Marketing, 20*, 277–319.

Hulland, J. (1999). Use of partial least squares (PLS) in strategic management research: A review of four recent studies. *Strategic Management Journal, 20*(2), 195–204.

Khanal, L. (2016). Awareness of knowledge management in Nepalese financial institutions. *Journal of Advanced Academic Research, 3*(3), 76–88.

Khanal, L., & Poudel, S.R. (2017). Knowledge management, employee satisfaction and performance: Empirical evidence from Nepal. *Saudi Journal of Business and Management Studies, 2*(2), 82–91.

Kline, R.B. (2011). *Principles and practice of structural equation modeling* (3rd ed.). New York: The Guilford Press.

Kianto, A., Vanhala, M., & Heilmann, P. (2018). The impact of knowledge management on job satisfaction. *Journal of Knowledge Management, 20*(4), 621–636. Doi: https://doi.org/10.1108/JKM-10-2015-0398

Lawal, W.O., Agboola, I.O., Aderibigbe, N.A., Owolabi, K.A., & Bakare, O.D. (2014). Knowledge sharing among academic staff in Nigerian University of Agriculture: A survey. *International Journal of Information Library & Society, 3*(1), 25–32.

Lawson, S. (2003). *Examining the relationship between organizational culture and knowledge management* (Doctoral dissertation). Nova Southeastern University.

Lee, H., & Choi, B. (2003). Knowledge management enablers, processes, and organizational performance: An integrative view and empirical examination. *Journal of Management Information Systems, 20*(1), 179–228.

Malhotra, N.K., & Birks, D.F. (2006). *Marketing research: An applied approach* (2nd European ed.). Harlow: Financial Times, Prentice Hall.

Malhotra, Y. (1998). Deciphering the knowledge management type. *Journal of Quality and Participation, 21*(4), 58–60.

Misener, T.R., Haddock, K.S., Gleaton, J.U., & Ajamieh, A.R. (1996). Toward an international measure of job satisfaction. *Nursing Research,* 45, 87– 91.

Nonaka, I., & Takeuchi, H. (1995). *The knowledge-creating company: How Japanese companies create the dynamics of innovation*. New York: Oxford University Press.

Nonaka, I., Toyama, R., & Konno, N. (2000). SECI, Ba and leadership: A unified model of dynamic knowledge creation. *Long Range Planning, 33*(1), 5–34.

Oshagbemi, T. (2000). Gender differences in the job satisfaction of university teachers. *Women in Management Review, 15*(7), 331–343.

Popa, I., Stefan, S.C., Cristina, M.C., & Cicea, C. (2018). Research regarding the influence of knowledge management practices on employee satisfaction in the Romanian healthcare system. *Amfiteatru Economic Journal, 20*(49), 553–566. Doi: https://doi.org/10.1108/JKM-10-2015-0398

Rana, G. (2010). Knowledge management and e-learning activities in the 21st century to attain competitive advantage. *Advances in Management, 3*(5), 54–56.

Rana, G., & Goel, A.K. (2017). Knowledge management process at BHEL: A case study. *International Journal of Knowledge Management Studies, 8*(1/2), 115–130.

Rana, G., Rastogi, R., & Garg, P. (2016). Work values and its impact on managerial effectiveness: A relationship in Indian context. *Vision, 20*(4), 300–311. Doi: 10.1177/0972262916668713

Rana, G., & Sharma, R. (2019). Assessing impact of employer branding on job engagement: A study of banking sector. *Emerging Economy Studies, 5*(1), 7–21.

Rasula, J., Vuksic, V.B., & Stemberger, M.I. (2012). The impact of knowledge management on organizational performance. *Economic and Business Review for Central and South-Eastern Europe, 14*(2), 147–168.

Sabherwal, R., & Sabherwal, S. (2005). Knowledge management using information technology: Determinants of short-term impact on firm value. *Decision Sciences, 36*(4), 531–567.

Saunders, M., Lewis, P., & Thornhill, A. (2003). *Research methods for business students*. Harlow: Pearson Education.

Scharmer, C.O. (2001). Self-transcending knowledge: Sensing and organizing around emerging opportunities. *Journal of Knowledge Management, 5*(2), 137–151.

Schultz, T. (1961). Investment in human capital. *The American Economic Review, 51*(1), 1–17.

Senge, P. (1997). Sharing knowledge. *Executive Excellence, 15*(6), 11–12.

Smith, F.J. (1976). The index of organizational reactions. *Catalog of Selected Documents in Psychology, 6*, 4–55.

Soo, C., Devinney, T., Midgley, D., & Deering, A. (2002). Knowledge management: Philosophy, processes and pitfalls. *California Management Review*, 44(4), 129–150.

Spector, P.E. (1985). Measurement of human service staff satisfaction: Development of the job satisfaction survey. *American Journal of Community Psychology, 13*(6), 693–713.

Tajali, M., Farahani, A., & Baharvand, M. (2014). Relationship between knowledge management with employees' performance and innovation. *Kuwait Chapter of Arabian Journal of Business and Management Review, 3*(11), 59–63.

Teece, D.J., Pisano, G., & Schuen, A. (1997). Dynamic capabilities and strategic management. *Strategic Management Journal, 18*(7), 509–533. Retrieved from https://josephmahoney.web.illinois.edu/BA545_Fall%202019/Teece,%20Pisano%20and%20Shuen%20(1997).pdf

Thakur, V., & Sinha, S. (2013). Knowledge management in Indian perspective. *The SIJ Transactions on Industrial, Financial, & Business Management (IFBM), 1*(1), 7–12.

Vinzi, V.E., Chin, W.W., Henseler, J., & Wang, H. (Eds.). (2010). *Handbook of partial least squares: Concepts, methods and applications*. New York: Springer.

Von Krogh, G. (1998). Care in knowledge creation. *California Management Review, 40*(3), 133–154.

Weiss, H.M., & Cropanzano, R. (1996). Affective events theory: A theoretical discussion of the structure, causes and consequences of affective experiences at work. *Research in Organizational Behavior*, 18, 1–74.

Zargar, E., & Rezaee, M. (2013). The study of knowledge management effect on performance rate of employees. *European Online Journal of Natural and Social Sciences,* 2(3), 3061–3066.

Zikmund, W.G. (2000). *Business research methods*. Fort Worth, TX: The Dryden Press.

9 Strengthening Employer Branding with Corporate Social Responsibility

Jeevesh Sharma, Suhasini Verma, and Shweta Taluka

INTRODUCTION

The branding word is very significant in itself because it puts impression in the customers' mind. Brand explains the position of a business and its product in the market among competitors. In general, we all are familiar with the term "branding" of the product and its importance only but the term "employer branding" is not so familiar among the customers and employees. So, it is very important to explore this term and find out its benefits. These will enhance its understandability, and will lead management and policymaker to come up with innovative strategies and implement them properly to gain more advantage from it. The core development and success of any business depend upon their operation, work culture, product specification, marketing, and advertising, reputation or company image, ethical and philanthropic of company's activities. Building and sustaining culture is a challenge for the organizations (Rana & Sharma 2017). These are the more considerable factors that impact the brand image of the company, among them the most ongoing process is the branding of products and services to consumers. Besides developing and advertising a product, employer branding also plays a crucial in the success of the business. As we all know the human resource of any company is the backbone of companies' operations. For best performance, the companies need talented candidates in their business, and these well-qualified, talented, enthusiastic, and committed employees are the requirement of any company (Sharma et al., 2018). Weak talent management can face severe risk in a talent war and it may result in high recruitment costs as essential to pay costs to poor reputation and encourage candidates to work for them despite the risk of problems. The higher the EB, the higher a company's potential to be a focus for, hiring, and maintaining more talented and committed candidates. Along with employer branding, the ethical behavior of a company also impacts its image (Rana, Rastogi, & Garg, 2016). It has been said that companies that are more responsible toward society or community are more considered by customers and investors as compared to those that are less responsible. In most studies, we learn CSR activities and company image or reputation, CSR, and good governance but a few studies are focused on CSR and EB. So this chapter explains in detail the key concept on and the relation of corporate social responsibility (CSR) and employer branding (EB). How corporate social responsibilities affect a company's attractiveness as an employer.

The remainder of the chapter discusses each aspect of EB and CSR individually. The first section of the chapter explains the EB and its strategies and importance of EB and CSR its meaning, CSR pillars, and major CSR stakeholders. Further, it explains the relationship between EB and CSR. Followed by the significance of EB with CSR and theories associated with it. The last section is followed by a conclusion and references.

EMPLOYER BRANDING

Meaning and Definition

At London Business School in 1996, Simon Barrow and Tim Ambler came with the term "employer branding" and defined it as "Employer branding is the package of functional, economic and psychological benefits provided by employment and identified with the employing company" (Ambler & Barrow, 1996, p.187). An enhanced company's EB creates competitive advantage which results in increasing recruitment and retains a more talented and committed candidate. Best EB seeks to best talent management and these results in increased company productivity and employee performance. EB is the unique aspect of a brand differentiates product manufacture and service providers from its competitors. Therefore, for a company, the employer as a brand represents an intangible asset. It not only refers to a unit of the category of employees but it also presents to a group of higher authorities (management).

EB can be categorized into internal and external branding:

Internal Employer Branding

Internal EB is focused on current employees. How to convince present employees to continue with existing employers and serve their duty with full responsibility? Proper working conditions, fringe benefits, periodically appraisals, bonuses, wages and salaries, training and development, bright future opportunity, personal growth, safety, and security. The entire factors influence the decision of the employees while working with the current employer. So it must be considered by the top level of management to boost internal EB among current personnel.

External Employer Branding

External branding of employer image includes making a good picture of the internal employer in front of jobseekers and key stakeholders. It comprises various plans to promote the internal goodness to external interest groups like internal working culture, perks, benefits, financial and non-financial incentives, future opportunities, etc. To accomplish talented and committed candidates the management will design creative career pages, hassle-free recruitment process, disclose more in a job advertisement, describing all benefits that would be offered to employees in detail, and be fixed with their promises.

Employer Branding and Employee Gratification

The above section discussed the meaning and definition of employee branding and explained the two constitutes of EB. It marks that EB is all about the image of an employer in the market and observed as a major factor among various factors while choosing a specific employer by job-seekers. Greater the employees' gratification of higher EB. Prior research had attempted to analyze whether employee satisfaction enhances the EB, and to the more extended researcher marked the positive relationship between them. Kaur and Syal (2017) explored the employer attraction dimensions to impact employer satisfaction and concluded that personnel-related principles, organizational development & culture principles, and personnel motivational principles of EB had a significant impact on employer satisfaction. Employee satisfaction directs the employees to decide on stay or join the other organizations (Sharma et al., 2019) Resultant to this attraction and retention of employees are affected by EB. Sokro (2012) remarked the evidence that aspect of the EB process in business operation impacts the decision-making process of existing employees and potential employees. And concluded that employers' favorable working atmosphere affects the mindset of employees through the survey of 87 employees.

The conclusion comes with is discussion is that the EB as a whole affect by the employees perspective and hence it must be considered in management planning to boost their existing employees with a comfortable working environment so that they can do their job with full energy and spread positive information of the organization in the market which may result in attracting talented human resources.

Importance of EB

EB awakes the potential and existing candidate for the company's brand. EB is the process by which employees frame a brand image and get motivated to draw an image of customers and other organizational elements. Therefore EB marks very important organizational constitutes, the following are the importance of EB.

Importance of Employer Branding

Sustainable human resource	* Highly competitive advantage
Stronger PR tool	* Favorable work culture
Increased shareholder value	* Talented human resource
Employee engagement	* Increase corporate identity

Fostering a relationship between an employer and employees

Strategic Activities of EB

Employer Value Propositions (EPV)

In 2005, Minchington coined this term. EPV can be defined as a group of association and offerings provided by a company reciprocally for the skills, capabilities, and experiences an employee renders while doing a job in an organization. It acts as a powerhouse of the employees which feels they are committed to the organization and doing a job as a pleasure. These senses of belongingness enhance the good working environment within the organization and employees enjoy their work. These competitive advantages of the company catch the attention of more potential human capital to the organization. Therefore, the employer value proposition presents the internal and external brand image of a company. Internal brand image includes the current employees and external brand image includes the major stakeholders. This would increase the value brand of the company and must configure with the present workers.

Linking with Campus

A socially responsible impression of the company cannot be ignored these days. And social responsibility initiatives are considered as a leading factor in the uplifting of the employer brand. Linking with campus can be understood as a collaboration of companies with the leading institutes. In recent past, companies are looking for initiatives to connect with the educational institute to pool the right talent for the organization. Now a day this platform of communication and interaction with future employees are serving a better opportunity to both candidate and employer to explore more things connected to both expectations.

Motivational Factors

Brand value is an intangible asset of companies, as it motivates both employers and employees towards the success of companies. To increase the branding of the employer the management is recognizing more and more motivational factors to influence the behavior of all levels of management. Motivational factors may include informative career page of company's website, culture and corporate values, socio-economic activities, communication of company's ethical initiative, the comfortable process of recruitment, career advancement opportunity, better induction of employees, job security, attractive wages, and salaries, promising nature of companies' management, etc. All these must be enforcing the existing and potential workers to work with the organization and achieve their desires. These motivating factors contribute to the development of EB and assist a worker to stay long with the organization.

Creative Marketing Team

Branding of a company is itself a creative activity performed by the marketing department. Branding initiatives try to establish a popular brand in consumers' minds because it increases their competitive advantages. Based on this branding of companies personnel is very unique strategies of management. It reflects the organizational working environment. CSR reflects all aspects of operation within a firm because of the need to consider the need to constitute groups. To make all these in public the creative marketing team is comprised so that more and more talented workforce can be attracted to organization, and give their best performance. Many points can be considered while planning for developing and maintains the EB like small interactive videos on social platforms, employee welfare focused advertisement, disclosing, and updating all societal activities on the company's website. All these initiatives may contribute to creating fascination among candidates and employees.

Feedback and Research

To improve the branding of a company the management can perform the feedback activities form the existing workforce working in an organization. This type of initiative form the company side can make the management know the loopholes existing in the functionality of a business. At the same time, it will also help the management to know the problems are confronting with current employees. This would lead the manager to research on the identified problem and initiate some plans to overcome this situation. All these activities would lead the company to explore the need and want current and prospective employees expected to form the organization. Resultant to this company would determine some attributes which would assist the manager to pool more and more talented human resource to an organization.

Statistics

To maintain and enhance the brand image of the firm as a good employer compared to other organizations, the management should maintain statistics on the performance of the firm in the market. This would help in tracking the position of a firm in the market and would help in making more strategic planning towards the enhancement of EB. By exercising these activities the major factor leading to better EB can be analyzed like brand awareness, satisfaction level of employees, an expectation of current and potential employees, and other company's initiative for their employees, etc.

CORPORATE SOCIAL RESPONSIBILITY

Meaning and Definition

CSR is the mainstream of business operations, as sustainable development is considered a major constraint of the company's business model. Because of this positive impact can be observed on socio-economic and environmental factors. Social related practices are the mode of the business to return to society what they had consumed in the process of business functioning. Under CSR the companies are developing a project almost covering all areas for the betterment of society like education, health & sanitation, rural development, promoting gender equality, promoting girl's education and scholarships, donating in government fund schemes, etc. So earlier it was voluntary for the company to spend on CSR activities but Company Act, 2013 institutionalized it and mentioned some specification in CSR law regarding it.

Social responsibility of business contents that business is responsible for the organization itself and also to other interest groups with which they interact. The interest groups include employees, customers, suppliers, government, society, environment, and investors. Positive relation among these group and organization results in increasing company value.

Company Act 2013

After the introduction of the Company's Act 2013, it has become a mandatory part of business operation and came into effect from April 2014 (Gazette of MCA). Company's Act 2013was introduced with some provisions related to CSR report. Companies that are operating in India, holding or subsidiary companies and foreign companies are covered under this law. As per law (Section 135 (1)) the companies have a net worth of Rs.500 crore or more; or turnover of Rs. 1000 crore or more; net profit of Rs. 5 crores or more are obliged to spend 2%of their average net of last three towards the community development, environment protection, and governance. Rules 8 & 9 of section 135 instruct for CSR reporting by an annual report through publishing annual CSR report as a separate section and same content on a website.

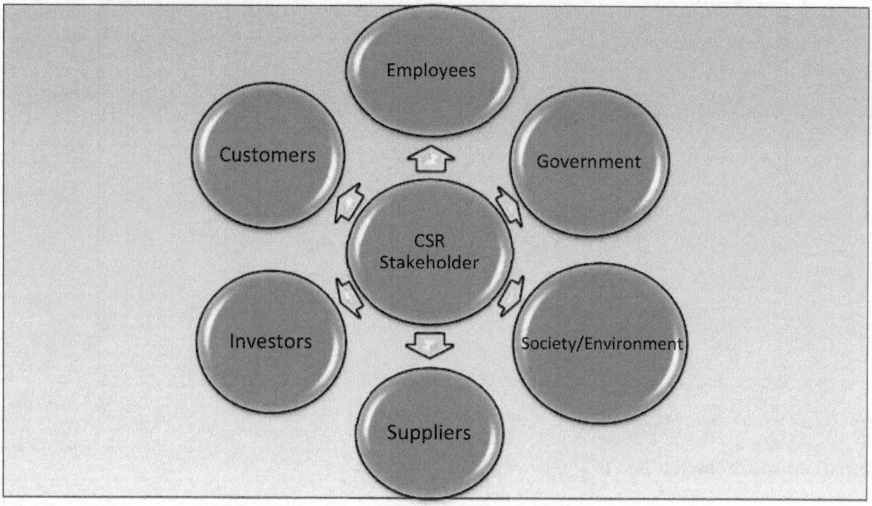

FIGURE 9.1 Stakeholders of corporate social responsibility.
(Source: Authors' own work.)

Apart from this Schedule VII of CSR law identified the 12 broad areas of CSR expenditure but companies are not limited to it. Based on these, companies are spending 2% (CSR fund) or more in these areas. Areas that are covered under Schedule VII are education, health, and sanitation, a donation to the relief fund and government agencies, rural development, promoting sports, eradication of poverty, etc.

CSR Pillars

CSR is driven by philanthropy. Triple bottom line defines CSR in 3Ps, that is, People, Profit, and Planet. Companies are serving all three in business operations. People stands for "Social," Profit stands for "Financial." And Planet stands for "Environment" responsibilities. Currently, CSR is a business strategy, because of performance consideration and stakeholder pressure. The social behavior of business is analyzed by all the stakeholders of companies like existing and potential customers, suppliers, government, shareholders, and employees. There are three pillars of CSR that is, economic, ethical, and legal, and almost all companies are working for all three pillars. It is a consideration of activities undertaken by the company to present the ethical obligation towards society.

Figure 9.2 presents the pillars of CSR. The three pillars are People, Profit, and planet. The company's responsibility toward people means the act of management toward the betterment of employees, customers, and society. The responsibility of the planet means initiating environmental behavior like a plantation, uses of advanced technology to reduce carbon profit, uses of natural resources, etc. And last but not least along with the responsibility of people and planet company work to earn

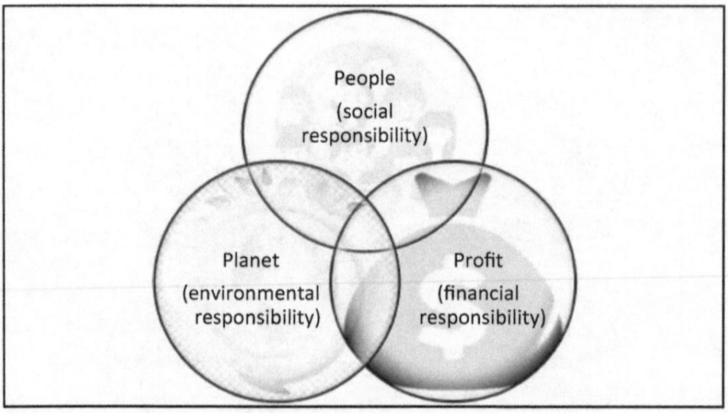

FIGURE 9.2 CSR pillars.
(Source: Authors' own work.)

profit for stakeholders those are associated with an organization. So, the impressive contribution of the company in these three pillars leads to a better image of the firm in the market and thus gains the name and fame and leads to an increase in the branding of the employer.

Corporate Social Responsibilities toward Various Stakeholders

Stakeholders	Concerned Activities
Environment	Lowering carbon emission, less depletion of natural resources, Use of renewable energy, following ISO standards, adopting advanced technology to reduce GHG emission, and proper waste management.
Society or Community	Promoting education, health, and sanitation camp, reducing gender inequality, developing rural areas, contributing to government funds or schemes, and welfare of the weak class of society.
Business	Green building, disclosing all relevant information to customers, producing recycled material, following legal rules and regulations. Utilize GMO (genetically modified organisms) ingredients for products, and Contribute 2% of profit for society upliftment
Employees	Retaining work–life balance, developing stronger relationships among employees and employer, developing an advanced training program, periodically appreciation, bonuses, safety and security, personal growth, financial and nonfinancial perks, proper leadership and motivational factors, promoting welfare program, Grievance cell, and protecting human rights.

(Source: Authors' own work.)

Corporate Social Responsibility and Branding

Before moving ahead first we should discuss how CSR practices are influencing the branding or brand equity of a firm. In this arena, numerous studies are undertaken to test the relationship between the firms' social responsibility and brand equity. As earlier we had discussed all the CSR ethical and legal obligations, we can conclude how CSR is becoming the most significant part of the business operation. Through the CSR activities company serve the society against what they borrowed from them. But CSR activities do not mean doing ethical initiatives towards the society and planet only it also includes the welfare and development of their workforce working in an organization. Employees' perception of an employer can be influenced by various firms' behavior. It can include both external and internal factors; internal factors count the company's HR policy, working culture, safety & security, management attitude, future perspective, etc. External factors are the inclusion of former employee's experience, product quality, consumer's perception, social activities of firms, etc. Yang and Basile (2019) analyzed the significant impact of CSR investment on the company's brand equity and concluded that poor quality of product weakens the relationship between employee-oriented CSR and brand equity. Rahman, Rodriguez-Serrano, and Lambkin (2019) also tested the relationship between corporate brand equity and companies' financial performance with CSR as a moderating variable and found the positive impact of CSR as a moderating variable between these two variables.

This literature proved that the attitude of employees does influence with company's CSR activities toward their both current and future employee. So nowadays companies are very attentive towards this issue and trying to develop various means and modes to satisfy their employees and made them comfortable with the business operation. It clearly marks the evidence of CSR and brand equity, brand equity increases EB.

EMPLOYER BRANDING AND CORPORATE SOCIAL RESPONSIBILITY

Concept

In an employer, branding evaluation is done based on communication, HRM, and marketing parameters. EB is a continuous process where information is collected from the existing employees on the efficiency of the recruitment process, retention rates, layoffs, and the company's external brand recognition. Other than this employee engagement, performance, and satisfaction are also considered important factors in the branding process.

CSR is also an ethical and legal performance of the company, where companies are willing to shell out extra and lay additional effort in business operation to satisfy and beneficial for the community and stakeholders. The performance of the company does impact the stakeholder but it also influences the workforce. Companies' ethical activities are not only observed by the consumers but also by the existing and potential candidates. It may not directly impact but indirectly it impacts personnel management. So the next section of the chapter explains to what degree EB and CSR are related and why it should be focused by the companies.

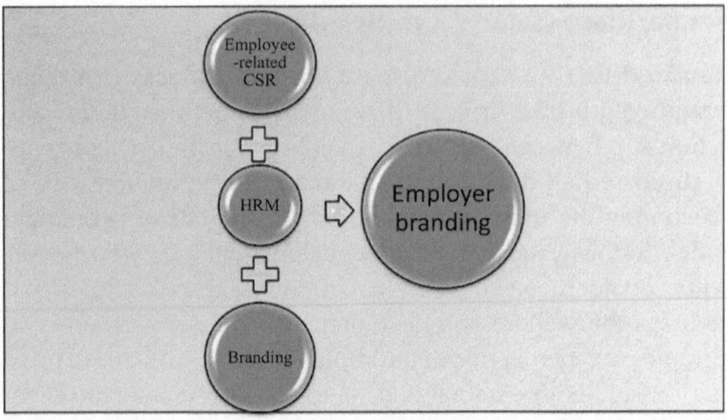

FIGURE 9.3 Conceptualization of employer branding.
(Source: Authors' own work.)

The above figure shows the conceptualization of employee-related CSR, HRM, and branding which bestow to the enlargement of EB. The above figure depicts that good CSR tactics of human resources increase the brand of the company and lastly it increase EB. EB can be differentiated between internal and external. Internal means existing employees related CSR like training and development, safety and security measures, monetary and non-monetary benefits, etc. External branding means the potential workforce like the recruitment process, creative marketing aids, employee protection, good & safe working condition, and fair wages and salaries.

Relationship between EB and CSR

EB is meant to be a critical part of CSR-oriented strategies. Sustainable human resource management and branding and CSR should be deliberately connected. EB is defined as financial, economical, and psychological benefits. Thus, the success of EB depends upon the strategic planning of all these functions. The GRI classification considered numerous HRM-related CSR indicators that are significant to the workforce and branding of the employer. The mentioned classifications are the inclusion of employment-oriented indicators such as financial compensations and setting up of work concerning work–life balance, labor-management, health and safety, training and education, and diversity and equality (Global Reporting Initiative [GRI], 2015). When a business operation behaves as a responsible entity towards workers and the community or society, the personnel not only experience valued but also get a chance to involve yourself contributing in good cause, which builds the trust within the employees is strengthened through positive attitude for an employer.

A strategic link exists between HRM, CSR, and branding, which requires an everlasting assurance from the workforce management, form apex level of management to every level of employees. Therefore, the function of HRM is imperative and can assist in the inclusion of corporate principles in the process of planning. The CSR information is should be displayed tactically on the company's website which encourages the stakeholders and potential employees to get engaged with the firm with their desired brand image. Thus, the most significant CSR stakeholder is the employees of a firm. This chapter also explains how communication patterns can also influence the perception of the workforce within the organization and outside the organization. Many researchers have done on this topic and they had concluded that the companies on their website disclose the content about corporate environment, recruitment, and job profession. To promote or strengthening EB through CSR the company categorizes the employee-related CSR activities into some indicators like recruitment-related, workforce-related, health and sanitation, improvement in education practices, safety and security, training and development, equality, and cause-related CSR actions.

There are many challenges faced by the organization within itself while hiring their target audiences like finding it difficult to attract the right skilled candidate for the position, trouble in retaining talented human capital, and high experience turnover rate. Thus, all these challenges direct the company to do exercise on it by reviewing the present organizational strategy. Some of the steps can be finding out expectations of the target audience, what they value to the company, what tools are they using to find a job, etc. Therefore to address these issues CSR activities would help in retaining and strengthening employee relations. Management can increase the scope of CSR disclosure in the job advertisement; in the online job recruitment portal, CSR information should be present easily under the section of company overview, offering opportunities to the employees' engagement with the company's CSR activities, etc.

In 2017, Deloitte studied and published on EB report on 77,000 individuals from 29 different countries and presented that 56% of young millennials are not interested to do jobs with those employers that are less focused on their values or behavior and 49% refused to work with the unethical organization.

Significance of EB and CSR

In previous sections, we learned the concept and relationship between EB and CSR. This section discusses the significance of EB and CSR.

Increase Organizational Social Behavior and Strengthen the Employer–Employee Relation

The company's social behavior increases the trust within the employees that "if their firm is doing right things" then "they are also contributing in right things." When companies are initiating to the best practice of CSR the employees feel more engaged in cooperative behavior towards their peer groups and organization. This results in encouraging a good relationship between employers and employees.

Boost Employee Identity with the Organization

The social responsibility of a firm may be more important than financial success. The socially responsible business gives a great experience to the employees with a sense of identity, where they are working. Employees do respect their company's activities and are found to be engaged with community welfare and development. Companies' best implementation of CSR practices boosts the employees' identity in the market as it increases the close relationship within organizations.

Enhance Employee Retention and Organizational Assurance

A sense of positivity about their organization's CSR actions has been resultant into increase employee's intension to continue with their existing employer. Organizational assurance in CSR practices also increases the employee interest with their current employer. The obligation includes a positive attitude, inclusion to what extent employees like their organization, sacrificing personal interest for the business, and vision their future and success joined to organization achievement. These entire scenarios enhance and improve employee retention.

Develop Employee Engagement and Performance

When an employee feels good with the company's CSR involvement, this increases employee engagement with organizational operations increases the performance of employees which results in increasing the success rate of the organization.

More Appealing Company Culture to Prospective Employees

With the increase of employee engagement and also the company's CSR commitment makes the company more appealing to prospective employees and applicants. Earlier millennial always look forward to work with the highly rated organization, engaging in CSR activities would help companies to attract more and more talented human resources over other companies.

The deliberate executions of CSR initiatives towards the employees not only create social consciousness among the personnel but also create a sense of adherent towards the organization and put on to the organization's positive brand image in the minds of the existing workforce and the community.

Facts and Figures

Randstad Employer Brand research (REBR)[1] is a body that researches every aspect which contributes to the enhancement of EB and ranked the companies based on their performance as the best employer. REBR 2020 has ranked the top Ten Indian Company as the best employer and ranked Microsoft India a technology massive company first as best EB in their survey, second to Samsung India, and on the third position is Amazon India. Other companies rank are Infosys at fourth, Mercedes–Benz in fifth, Sony in the sixth, IBM in seventh, Dell technologies in eight, ITC at ninth, and TCS was at tenth rank. As in the survey, every aspect related to EB is analyzed, so Microsoft India Company grabs this position based on financial health, strong reputation, and implementation of advanced technologies in the workplace.[2] Whereas in the year 2019 survey REBR released the top 10 best employers where Amazon India was ranked first as the most attractive employer brand in India and Microsoft India was a runner. Sony India was in the third position, Mercedes–Benz in the fourth position, IBM in the fifth position, L&T sixth position, Nestle in the seventh position, Infosys on eight positions, Samsung on ninth position, and last ranker was

Dell. REBR 2019 survey concluded that salary and employee-oriented benefits, job security, and work–life balance are the most considered indicators while choosing employers by job seekers in India. But in the REBR 2020 survey, they announced work–life balance as the most attractive indicator observed by the job-seekers.

These facts and figures depict the true picture of the employee expectation towards the brand of employer. This shows how the choice of prospective employees to employers is varying day by day and the presence of employee mind while seeking an employer.

Signaling Theory

CSR as a mandatory action of firms so disclosing CSR information to stakeholders is the primary responsibility of the firm. Signaling means providing signals to concerned interest groups regarding certain practices. In the context of CSR, signaling theory describes how and what companies adopting various modes to convey information regarding their capabilities to the stakeholders. In recent past years, signaling theory is making space in strategic decisions of management to understand how information is communicated among parties. The observation of signaling theory enables the external to the firms to watch the extent of firms' morale and tactics such as CSR behavior to the talent market. Prior study has discussed CSR performance with signaling theory to elucidate the potential return companies will obtain from adopting and remarkable in socially liable practices. So the firm's societal actions serve up as a signal of job culture or environment at the firm. Therefore, existing personnel and potential personnel employ CSR practices as a signal to reduce information asymmetry and come to decide whether to continue with the existing employer and to apply for a job.

Social Identity Theory

Social identity theory explains the reason how a person perceives their belongingness to a specific group and takes some actions to support this group. It means it is directly connected to the organizational affiliation, which includes the values of the firm which is portrayed for the public. Companies with higher CSR values generate higher value as compared to low CSR values. Job seekers analyze the values of companies with their pre-determined values before applying for a job. Therefore these connections proved that companies having greater labor or employee-related CSR performance are most admired by the talent market.

Thus, these two theories show the association of how firms' CSR communication can impact the firm image. Through the lens of these two theories, we can conclude that communication strategies are also equally important for a firm's image, which is observed by both existing and potential employees.

CONCLUSION

With the increase of corporate ethical behavior, the manpower in the workplace is also efficiently performing their responsibility, and resultants to this potential candidate are interested to be recruited in such firms. CSR plays a crucial role in developing and maintains the reputation of a firm in public image and hence increases the brand image among major stakeholders. The companies are appointing employees as volunteers in the company's CSR activities. As human resource is the backbone of the organization, for the success of business operation talented candidates are the requirement of the organization. Here, EB is the major source for attracting talent market to the organization, and this can be raised through a major role act by a company's ethical behavior toward their employee and society. Dawn and Biswas (2010) concluded that employee values impact EB and those values are social, interest, development, and economic values. The study employed the theories also to understand the depth of the relationship between employer and CSR. Signaling theory provides the platform to know about the company and social identity theory supports the perception of employees toward the company. Thus, the managerial and theoretical implications of the study can be elaborated through exploring all the aspects related to brand CSR and EB. The findings of the chapter will enable the manager and the policymaker to frame and implement these strategies in their forum so that more and more talented human capital can be attracted in organization operation and will gain competitive advantage among all rivalries. The theoretical contribution of the study concludes that the communication of CSR activities also impacts the performance of the firm. It can be understood by the signaling theory, which states that, if the communication channel is not right then the activities of the firm cannot be approached by the targeted population. These strategies contribute to the strengthening of the EB of the organization through CSR. The limitation of the present study is that it is done based on reviewing literature or secondary sources. So it directs future research, which can be done by taking a sample of companies and conducting a survey of company's employees by performing primary data analysis. The results of this study will surely present the true picture of EB among employees.

NOTES

1 www.livemint.com/companies/news/microsoft-india-most-attractive-employer-brand-survey-11595920573946.html

2 www.livemint.com/companies/news/amazon-india-most-attractive-employer-brand-says-survey-here-are-others-on-the-list-1560765804676.html

REFERENCES

Ambler, T., &Barrow, S. (1996). The employer brand. *Journal of Brand Management*, 4, 185–206.

Backhaus, K., & Tikoo, S. (2004). Conceptualizing and researching employer branding. *Career Development International*, 9(5), 501–517.

Benitez, J., Ruiz, L., Castillo, A., & Llorens, J. (2020). How corporate social responsibility activities influence employer reputation: The role of social media capability. *Decision Support Systems*, 129, 113–223.

Biswas, M. K., & Suar, D. (2016). Antecedents and consequences of employer branding. *Journal of Business Ethics*, 136(1), 57–72.

Bustamante, S. (2018). CSR, trust and the employer brand (No. 96). Working Papers of the Institute of Management Berlin at the Berlin School of Economics and Law (HWR Berlin).

Bustamante, S., & Brenninger, K. (2013). CSR and its potential role in employer branding an analysis of preferences of German graduates. In: Making the number of options grow. Contributions to the corporate responsibility research conference.

Carlini, J., Grace, D., France, C., & Lo Iacono, J. (2019). The corporate social responsibility (CSR) employer brand process: Integrative review and comprehensive model. *Journal of Marketing Management*, 35(1–2), 182–205.

Dawn, S. K. & Biswas, S. (2010). Employer branding: A new strategic dimension of Indian corporations. *Asian Journal of Management Research*, 21–33, ISSN 2229 – 3795.

Della Corte, V., Mangia, G., Micera, R., & Zamparelli, G. (2011). Strategic employer branding: The brand and image management as attractiveness for talented capital. *China-USA Business Review*, 10(12), 1231–1252.

Figurska, I., & Matuska, E. (2013). Employer branding as a human resources management strategy. *Human Resources Management & Ergonomics*, 7(2), 35–51.

Gehlo, R., & Patil, D. (2019). *Importance of corporate social responsibility in employer branding*. Allana Institute of Management Sciences, Pune, 9, 1–5.

Jakopovic, H. (2017). Employer branding through CSR and survey. *Economic and Social Development: Book of Proceedings*, 620–629.

Kádeková, Z., Savov, R., Košičiarová, I., & Valaskova, K. (2020). CSR activities and their impact on brand value in food enterprises in Slovakia based on foreign participation. *Sustainability*, 12(12), 4856.

Kashikar-Rao, M. (2014). Role of CSR in employer branding: Emerging paradigm. *Review of HRM*, 3, 188.

Kaur, J., & Syal, G. (2017). Determinative impact of employer attractiveness dimensions of employer branding on employee satisfaction in the banking industry in India. *Business Analyst*, 37(2), 129–144.

Kharisma, P. (2012). *The role of CSR in employer branding strategy: From legitimacy to organizational commitment*. Dijon: University of Bourgogne.

Klimkiewicz, K., & Oltra, V. (2017). Does CSR enhance employer attractiveness? The role of millennial job seekers' attitudes. *Corporate Social Responsibility and Environmental Management*, 24(5), 449–463.

Kucharska, W. (2020). Employee commitment matters for CSR practice, reputation and corporate brand performance—European model. *Sustainability*, 12(3), 940.

Lindholm, L. (2018). The use of corporate social responsibility in employer branding. In: Koporcic, N., Ivanova-Gongne, M., Nyström, A.-G. and Törnroos, J.-Å. (Ed.) *Developing Insights on Branding in the B2B Context*. Emerald Publishing Limited, pp. 73–93.

Prieto, A. B. T., Shin, H., Lee, Y., & Lee, C. W. (2020). Relationship among CSR initiatives and financial and non-financial corporate performance in the Ecuadorian banking environment. *Sustainability*, 12(4), 1621.

Puncheva-Michelotti, P., Hudson, S., & Jin, G. (2018). Employer branding and CSR communication in online recruitment advertising. *Business Horizons*, 61(4), 643–651.

Puncheva-Michelotti, P., Hudson, S., &Jin, G. (2018). Employer branding and CSR communication in online recruitment advertising. *Business Horizons*, 61(4), 643–651.

Rahman, M., Rodríguez-Serrano, M. Á., & Lambkin, M. (2019). Brand equity and firm performance: the complementary role of corporate social responsibility. *Journal of Brand Management*, 26(6), 691–704.

Rana, G., & Sharma, R. (2017). Organizational culture as a moderator of the human capital creation-effectiveness. *Global HRM Review*, 7(5), 31–37.

Rana, G., Rastogi, R., & Garg, P. (2016). Work values and its impact on managerial effectiveness: A relationship in Indian context. *Vision*, 22, 300–311.
Sharma, R., Jain, V., & Singh, S. P. (2018). The impact of employer branding on organizational commitment in Indian IT sector. *IOSR Journal of Business and Management*, 20(1), 49–54. https://doi.org/10.9790/487X-2001054954
Sharma R., Singh S. P., Rana G. (2019) Employer branding analytics and retention strategies for sustainable growth of organizations. In: Chahal H., Jyoti J., Wirtz J. (Eds.) *Understanding the Role of Business Analytics*. Singapore: Springer. https://doi.org/10.1007/978-981-13-1334-9_10.
Sokro, E. (2012). Impact of employer branding on employee attraction and retention. *European Journal of Business and Management*, 4(18), 164–173.
Verčič, A. T., & Ćorić, D. S. (2018). The relationship between reputation, employer branding and corporate social responsibility. *Public Relations Review*, 44(4), 444–452.
Yang, J., & Basile, K. (2019). The impact of corporate social responsibility on brand equity. *Marketing Intelligence & Planning*, 37(1), 2–17.

10 Enhancing Employee Happiness

Branding as an Employer of Choice

Rinki Dahiya

INTRODUCTION

People consider virtues as their highest aspirations. In the organisational context, Cameron et al. (2004) defined organisational virtuousness as "the good habits, actions and desires that are nourished, practiced, supported and disseminated, both at collective and individual levels." Due to lack of interest of researchers, studies on organisational virtuousness lagged behind in the organisational literature. However, with the emerging financial turmoil and moral downfalls across the business world, popular business communities and renowned business presses started to believe that nurturing workplace virtues, may improve organisational efficiency and individual betterment of employees (Dahiya & Rangnekar, 2020a; Cameron, 2010; Tsachouridi & Nikandrou, 2019). Further, positive organisational researchers stress that virtues need to be included in management and business research agenda (Dahiya & Rangnekar, 2018b; Rego et al., 2010). Nevertheless, some authors attempted to empirically examine the organisational virtuousness (Cameron et al., 2004; Lillius et al., 2008; Tsachouridi & Nikandrou, 2019). However, the manifestations of virtuousness in the organisational context remained under-developed theoretically in terms of full range of consequential organisational phenomena (Wright & Goodstein, 2007; Tsachouridi & Nikandrou, 2019). Therefore, the present study seeks to nourish this by investigating that how organisational virtuousness predicts the perceptions of employer branding, both directly and indirectly as a psychological mediating mechanism of employee happiness.

Happiness of employees is one of most crucial future research agenda for contemporary organisations (Dahiya & Rangnekar, 2018a). As organisations of today need to ensure the happiness and overall well-being of their employees if they are to remain engaged and productive. The "holy grail" of management research has signified that happy employees are high on productivity in contrast to unhappy employees. Researchers have highlighted that happiness will offer a better understanding of what drives employees to flourish by realising their latent abilities at work and contributing towards organisational success and competitive advantage. Also, a possible solution to the problem of attrition is to enhance employee value proposition and organisational attractiveness; and happiness is the key to achieve it. Regardless of all organisational related outcomes of happiness, little is known about the relationship between happiness and perceptions of employer branding. Organisations of today are striving for a virtuous and happy workforce to maintain their competitive edge by becoming a brand as an employer of choice (Dahiya & Rangnekar, 2019c; 2020b; Frenking, 2016; Newstead et al., 2019). Therefore, the present study aims to examine that whether enhancing employee happiness affects the perceptions of employer branding directly and indirectly psychological mechanism of happiness.

THEORETICAL BACKGROUND AND LITERATURE REVIEW

Linking Perceptions of Organisational Virtuousness with Employer Branding

Employees have a tendency to respond to their organisational associations by modifying their frames of mind in a reciprocating manner (Dahiya & Rangnekar; 2020c). Additionally, through organisational virtuousness employees may develop feelings of gratitude, seek for authoritative support and create social mental agreements with the organisation (Froh et al., 2008). In this manner, they develop higher sense of commitment towards the associated organisation (Newstead et al., 2019). Further, they develop feelings of doing meaningful work, and get their whole self (spiritual, emotional, mental and physical) dedicated to the organisation, and consider their assigned duties, and responsibility more than as a mere "job." This further leads to their strong connection with the organisation or brand (Milliman et al., 2003). Hence, the first hypothesis is as below:

H_1: Employees who experience better organisational virtuousness develop higher perceptions of employer branding.

Linking Employee Happiness and Perceptions of Employer Branding

Taking into account that perceptions of employer branding is not only an emotion based but also cognitive based bond of the employees towards their organisations, therefore, conceivably happiness at work is a potential predictor of employer branding (Frenking, 2016). In other words, if people experience happiness at workplace, they build up an emotional connection with the image of their organisation. The same has been empirically evidenced by contemporary researchers in the field, for example; Dahiya and Rangnekar (2019b), Fisher (2010) and Salas–Vallina et al. (2017). Furthermore, Lilius et al. (2008) also proposed that incessant workplace experiences that generate positive feelings may prompt wonderful enthusiastic relationship with the organisation. Also, such positive emotions accrue after some time build into strong associations with the employer brand.

As recommended by the broaden and build theory (Fredrickson, 1998), workplace happiness may motivate people to consider their work as meaningful, therefore leading them to regard their work as a "mission" instead of just a "job." Thusly tend them all the more emotionally appended to their associations and increasingly dedicated to improve not only their own performance but also the reputation of their employer brand. Further, Kuvaas (2006) also argued that positive emotions related to work may likewise make their job characteristically and intrinsically fulfilling, hence advancing their good perceptions about the employer brand. Henceforth, the present study aims to test the following hypothesis:

H_2: Employees who experience better happiness at work develop higher perceptions of employer branding.

Linking Perceptions of Organisational Virtuousness and Employee Happiness

Researchers argued that when individuals get exposed to virtuousness, they develop emotions such as zest, enthusiasm, empathy, zeal and love (Dahiya & Rangnekar, 2018a; 2018b). Also, virtuous contexts encourage high-quality connections which cultivate feelings of meaningful work, as a result stimulating further positive emotions (Tsachouridi & Nikandrou, 2019). Additionally, employees may start feeling not only as a valuable emotional but also as intellectual beings instead of just as a "human resource" of an organisation. Consequently, owing to the feelings of recognition, they develop gratitude towards the organisation; and this frame of mind may enhance their well-being (Fredrickson, 2001).

Dutton and Heaphy (2003) also highlighted that if the employees experiences organisational virtuousness, then they not only start developing positive images but also get attached to the virtuous actors of the organisation, and develop quality social bonds. Such premium connections at work support the employees to gratify their security and social needs, thereby contributing towards their higher happiness (Baker & Dutton, 2007). Also, broaden and build theory (Fridrickson, 2001) stated that when people work in a virtuous organisation, they tend to experience positive emotions. This further enhances the horizon of their thought action repositories and increases their problem-solving abilities creatively. Moreover, they deal with the work hassles in an open manner which lessens their frustrations, and encourages the employee happiness. Therefore, this study posits that:

> *H_3: Employees with better perceptions of organisational virtuousness experience higher happiness.*

Mediating Role of Employee Happiness in the Link between the Perceptions of Organisational Virtuousness and Employer Branding

If organisational virtuousness enhances employee happiness; and happiness positively affects employer branding, then the organisational virtuousness is likely to affect the perceptions of employer branding through the psychological mechanism of happiness. Consequently, the mediation process is possibly partial as virtuousness also influence perceptions of employer branding through the mechanism of happiness. The integrated relationship of these constructs can be understood with the background of affective events theory propounded by Weiss and Cropanzano (1996). Affective events are day-to-day organisational occurrences that stimulate emotional reactions of employees. However, the present research do not assess them openly but argue that the organisational virtuousness spark positive events such as being treated with compassion and/or courtesy which as a result evokes positive emotions. In such a way employees lead to experience positive emotions because such events support their needs, interests, goals, interests and well-being (Dahiya & Rangnekar, 2020d; Spector & Fox, 2002). Also, as stated by Lilius et al. (2008), the frequent pleasant associations accumulate with time and strengthen employee's perceptions of employer branding. Thus, hints given by the findings of the earlier studies mentioned above allow us to hypothesise as below and Figure 10.1 presents the hypothesised model:

> *H_4: Employee happiness mediates the relationship between perceptions of organisational virtuousness and employer branding.*

METHODS

Sample and Procedure

The participants were the employees working in North Indian manufacturing organisations majorly functioning in the area of automobile, infrastructure and power generation. The data was collected through survey method and convenience sampling was adopted. A sample of 600 respondents was targeted, of which 386 returned the

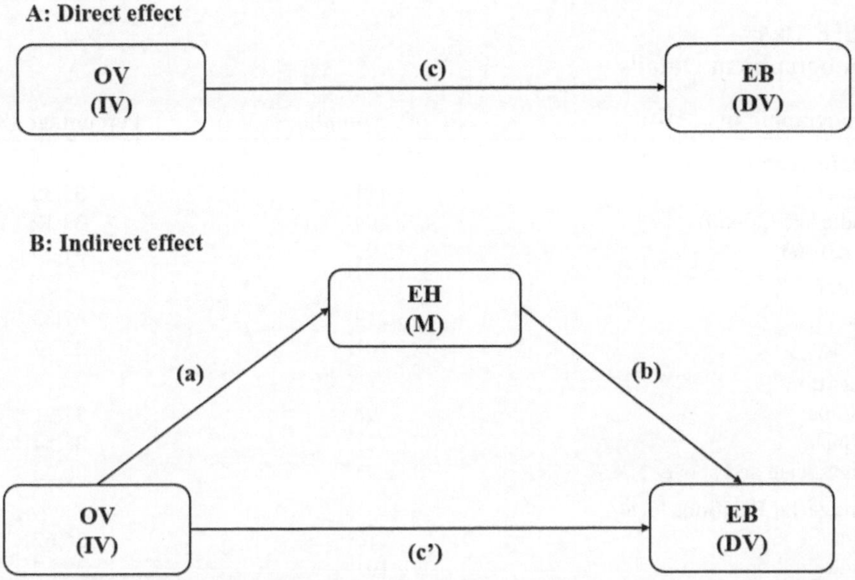

FIGURE 10.1 Hypothesised model.

Note: OV = perceptions of organisational virtuousness; EB = employer branding; EH = employee happiness.

(Source: Author's own work.)

questionnaire. After the initial screening, 314 responses were found to be suitable for data analysis. Table 10.1 presents the details of respondents.

Measures

Organisational Virtuousness. The employees' perceptions of organisational virtuousness were measured with the scale developed by Cameron et al. (2004). The measurement scale has 15 items with five sub dimensions (three items in each), namely, organisational compassion, organisational optimism, organisational integrity, organisational trust and organisational forgiveness. The responses were taken on a 7-point Likert-type scale ranging from 1 (false) to 7 (true). The illustrative items for organisational virtuousness are "A sense of profound purpose is associated with what we do here" and "This organisation demonstrates the highest levels of integrity." In the present study, the value of alpha for complete scale was 0.813.

Employee Happiness. The perceptions of employee happiness were measured with shorter version of Happiness at Work (HAW) Scale validated by Salas–Vallina and Alegre (2018). It has nine items which include three dimensions, namely, job satisfaction, engagement and affective organisational commitment. The responses were taken on a 7-point Likert-type scale wherein 1 stands for "strongly disagree"

TABLE 10.1
Demographical Details

Demographic (n = 314)	No. of respondents	Percentage (%)
Age (in years)		
Young (21–35)	111	35.35
Middle age (36–50)	104	33.12
Old (51–65)	99	31.53
Gender		
Male	213	67.83
Female	101	32.17
Education		
Diploma	98	31.21
Graduate	121	38.54
Postgraduate and above	95	30.25
Managerial Positions level		
Junior	115	36.62
Middle	101	32.17
Senior	98	31.21
Work Experience (in years)		
Less than 10	118	37.58
10–20	98	31.21
More than 20	98	31.21

and 7 stands for "strongly agree." The sample items are "At my job, I feel strong and vigorous" and "I would be very happy to spend the rest of my career with this organisation." The Cronbach's alpha for the present study is 0.857.

Perceptions of Employer Branding. The perceptions of employer branding was measured by adapting the scale developed by Hillebrandt and Ivens (2013). It has instrumental, symbolic framework and the sample item is "This organisation is one of the best employers to work for." The responses were taken on a 7-point Likert-type scale wherein 1 stands for "strongly disagree" and 7 stands for "strongly agree." The Cronbach's alpha for the present study is 0.825.

RESULTS

Hypotheses Testing

After the preliminary tests such as mean, standard deviation, correlation (see Table 10.2) and assumptions of regression analysis were tested such as multi collinearity, homoscedasticity, linearity, normality of data and the next was to test the hypothesis. In order to test the same, direct (effect of organisational virtuousness on happiness, organisational virtuousness on employer branding and effect of happiness on employer branding) and indirect effects (effect of organisational virtuousness on employer branding via happiness) were computed with hierarchical regression

TABLE 10.2
Descriptive Statistics

Variables	Mean	SD	OV	EH	EB
OV	4.84	1.05	***(0.813)***		
EH	5.02	1.56	0.224*	***(0.857)***	
EB	4.67	1.56	0.201***	0.237**	***(0.825)***

Note: SD = Standard Deviation; OV = organisational virtuousness; EH = employee happiness; EB = employee branding; the reliability coefficients (α) appear in bold and diagonal in parentheses; significance at $*p < 0.05$, $**p < 0.01$ and $***p < 0.001$.

TABLE 10.3
Results of Hierarchical Regression for Direct Effects

	Predictors	Employer Branding (DV) β		Happiness (DV) *B*	
		Step 1	Step 2	Step 1	Step 2
Step 1:	***Control Variables***				
	Age	0.221*	0.107	0.282*	0.101
	Gender	0.169*	0.109	0.221**	0.009
	Education level	0.111*	0.098	0.111**	0.090
	Managerial Positions	0.109*	0.080	0.109**	0.091
	Work Experience	0.211*	0.105	0.254*	0.019
Step 2:	***Predictor*** **(IV)**				
	OV		0.341**		0.296**
	ΔF	12.145**	36.294**	11.593**	31.762**
	R^2	0.119	0.241	0.105	0.215
	Adjusted R^2	0.114	0.237	0.100	0.213
	ΔR^2	—	0.122**	—	0.110**

Note: N = 432; IV = independent variables; DV = dependent variables; β = standardised beta coefficients are reported in the regression table; OV = organisational virtuousness; $**p < 0.01$; $*p < 0.05$.

analysis. Additionally, Baron and Kenny's (1986) recommendations were followed in testing the joint significance of indirect effect. First of all, direct effect of organisational virtuousness on employer branding was checked. In step 1, following the procedure adopted by Dahiya and Rangnekar (2019a) confounding variables were entered in block 1, and step 2 was followed by entering independent variable (organisational) in block 2. Similar steps were followed for testing the direct effect of organisational virtuousness on happiness (see Table 10.3 for results of hierarchical regression).

The results highlighted that organisational virtuousness accounts for 24.1% variance (ΔF (1, 426) = 36.294; adjusted $R^2 = 0.237$; $\Delta R^2 = 0.122$, $p < 0.01$) in the perceptions of employer branding. Also, the standardised coefficient beta ($\beta = 0.341$, $t = 3.904$, $p < 0.01$) revealed significant direct effect on happiness. Further, the direct

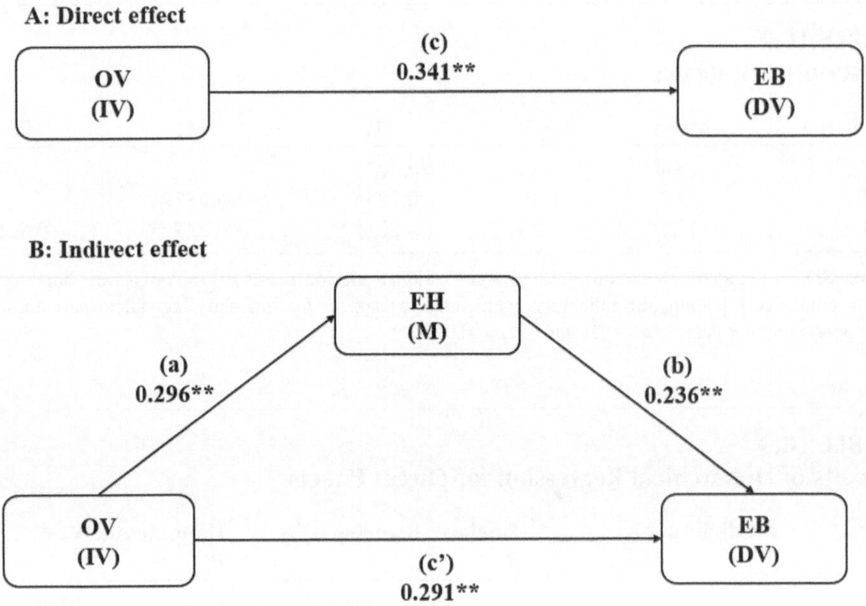

FIGURE 10.2 Results of hypothesised model.

Note: OV = perceptions of organisational virtuousness; EB = employer branding; EH = employee happiness. β = standardised beta coefficients are reported; $**p < 0.01$.

(Source: Author's own work.)

effect of organisational virtuousness on happiness was checked and results highlighted that organisational virtuousness accounts for 21.5% variance (ΔF (1, 426) = 31.762; adjusted $R^2 = 0.213$; $\Delta R^2 = 0.215$, $p < 0.01$) in happiness. Also, the standardised coefficient beta ($\beta = 0.296$, $t = 3.732$, $p < 0.01$) revealed significant direct effect on happiness. These results (direct effect) suggested that organisational virtuousness is a significant predictor of employer branding and employee happiness which support first and third hypothesis.

Further, the indirect effect of organisational virtuousness on employer branding via happiness was checked by conducting hierarchical multiple regression. The results presented in Table 10.4 indicated that organisational virtuousness and happiness collectively account for 32.3% variance (ΔF (2, 425) = 92.546; adjusted $R^2 = 0.319$; $\Delta R^2 = 0.210$, $p < 0.01$) in perceptions of employee branding. Also, the standardised coefficient beta for organisational virtuousness on employer branding was reduced with the presence of happiness in the regression model but remained significant ($\beta = 0.291$, $t = 3.712$, $p < 0.01$). Additionally, standardised coefficient beta for happiness on perceptions of employer branding was significant ($\beta = 0.236$, $t = 3.453$, $p < 0.01$). Therefore, supporting second hypothesis that employees who experience better happiness at work develop higher employer branding while for fourth hypothesis there was partial support for the intervening effect of employee

TABLE 10.4
Results of Mediation Analysis

	Predictors	Employer Branding (DV) β Step 1	Step 2
Step 1:	***Control Variables***		
	Age	0.221*	0.107
	Gender	0.169*	0.109
	Education Level	0.121*	0.101
	Managerial Positions	0.108**	0.090
	Work Experience	0.211*	0.105
Step 2:	***Predictor* (IV)**		
	Organisational Virtuousness		0.291**
	Employee Happiness		0.236**
	Δ *F*-Value	35.118**	92.546**
	R^2	0.113	0.323
	Adjusted R^2	0.111	0.319
	ΔR^2	—	0.210**

Note: N = 314; IV = independent variables; DV = dependent variables; β = standardised beta coefficient are reported in the regression table; **$p < 0.01$; *$p < 0.05$.

happiness in the association between organisational virtuousness and perceptions of employer branding.

Furthermore, following the recommendation of Baron and Kenny (1986), the significance of indirect effect (organisational virtuousness on employer branding via happiness) was tested with by calculating the joint significance of direct effects (a*b; a = effect of organisational virtuousness on happiness; b = effect of happiness on employer branding). As shown by the results and Figure10. 2, both paths "a" and "b" were significant and supported that a*b is also significant. Also, an alternative method to estimate the significance of indirect effect given by Sobel (1982) was followed by using Sobel's test. Results supported significant indirect effect [standardised indirect effect (a*b) = 0.061; Sobel SE = 0.028; Z value = 2.188; $p = 0.028$; $p < 0.05$] and also standardised indirect effect, that is, a portion of organisational virtuousness on employer branding due to happiness was 14.66% [portion of (X → Y due to M) = (c-c')/c). These results partially support the fourth hypothesis.

DISCUSSION, CONCLUSION AND IMPLICATIONS

The study aims to investigate the direct effect of organisational virtuousness on perceptions of employer branding along with indirect effect via employee happiness. The results indicated that organisational virtuousness and employee happiness significantly influence perceptions of employer branding and employee happiness partially mediates this relationship. As hypothesised employee who experience better organisational virtuousness develop higher perceptions of employer branding. Results are consistent with the Froh et al. (2008) and Newstead et al. (2019) that the organisational virtuousness is beyond the ethical courses and encourages psychological growth and development of employees. Also, in alignment with the findings of Milliman et al. (2003), association with the virtuous organisation generally enhances the feelings of gratitude in the employees towards their organisation and also tend to increase the value and attractiveness of organisation in their eyes. On the similar lines, in the present study, virtuousness of the organisation tends to enhance the perceptions of employer branding.

The findings of the second hypothesis highlighted that employees who experience happiness at work develop higher perceptions of employer branding. To be precise, employee who experiences virtuousness in the organisation such as forgiveness, trust, optimism and compassion forms an attachment and attraction with their organisation; they feel proud to identify themselves with such an organisation and may serve their responsibilities with more diligence (Singh et al., 2018). Also, consistent with the findings of Lilius et al. (2008), happy employees tend to form strong social bonds within and outside the organisations, find their work more meaningful and derive work satisfaction; this positivity roots from working in the virtuous organisation and tend to increase the perceptions of employer branding.

The results of the third hypothesis reveal that employees with better perceptions of organisational virtuousness experience higher happiness. Corroborating with broaden and build theory (Fredrickson, 2001) and aligned with the findings of Dutton and Heaphy (2003) and Baker and Dutton (2007) that experiencing virtues at work develop positive emotions, thereby contributing towards happiness of employees. The fourth hypothesis supported a partial mediating effect of happiness on the association between organisational virtuousness and perceptions of employer branding. This is similar with the Cameron et al.'s (2004) and Rego et al.'s (2010) study that people working in virtuous work environment develop positive perceptions about the employer brand and report not only higher positive cognitive evaluations of their job but also experience greater happiness.

Employees experience emotional events or encounter organisational level factors as salient outcomes of organisational virtuous that spark positive emotions in them for the organisation and strengthen their perceptions about the employer brand (Lilius et al. 2008). Furthermore, as stated by Herrbach (2006), certain virtuous features of the work environment develop positive moods and emotions through constant activation of employee's appetitive system and correspondingly develop notions of commitment, attractiveness of the organisation and perceptions regarding the employer brand.

The present research addresses the call from the positive organisational studies movement (e.g., Cameron et al., 2004; Wright & Goodstein, 2007)and future research agenda stressed by contemporary researchers in the organisational virtuousness area (e.g., Frenking, 2016; Newstead et al., 2019) that emphasise on the comprehensive understanding of the factors that not only develops but also sustains employees' positive mental states in organisations. In this way, employees who experience organisational virtues experience higher happiness and also greater perceptions of employee brand. The main conclusion is that organisations can promote a happy workforce if they encourage positive perceptions regarding the organisational virtues in their employees with a genuine and sustainable way. With this recognising, there is a potential favourable impact of organisational virtuousness and employees' happiness on positive perceptions of employer brand. As Dahiya and Rangnekar 2018a also pointed out that it is highly practical and reasonable for both management scholars and business executives to understand that employee happiness is not only a corporate "buzz" word but also a valuable tool to make the most of employee's potential for both personal and professional betterment.

LIMITATIONS AND FUTURE DIRECTIONS

The limitations of the findings are the following: first, the analysis was based on cross-sectional data, so in future longitudinal research design may be conducted for comprehensive understanding of the association between variables. Second, for data collection, a survey method was adopted; therefore, in future other methods such as daily diary data, expert observation and interview may be conducted to support the findings of the present study. Third, the association between organisational virtuousness and perceptions of employer branding can be studied with other variables, such as work engagement, emotional stability and affective commitment. Moreover, the moderating role of demographical variables, for example, gender and age, may also be considered in future.

REFERENCES

Baker, W., & Dutton, J. E. (2007). Enabling positive social capital in organizations. In J. E. Dutton & B. Ragins (Eds.), *Exploring positive relationships at work: Building a theoretical and research foundation* (pp. 325–345). Mahwah, NJ: Lawrence Erlbaum.

Baron, R. M., & Kenny, D. A. (1986). The moderator–mediator variable distinction in social psychological research: Conceptual, strategic, and statistical considerations. *Journal of Personality and Social Psychology*, *51*(6), 1173–1182.

Cameron, K. (2010). Five keys to flourishing in trying times. *Leader to Leader*, 55, 45–51.

Cameron, K. S., Bright, D., & Caza, A. (2004). Exploring the relationships between organizational virtuousness and performance. *American Behavioral Scientist*, *47* (6), 766–790.

Dahiya, R., & Rangnekar, S. (2018a). Employee happiness a valuable tool to drive organizations. In M. Yadav, S. K. Tridevi, A. Kumar, & S. Rangnekar (Eds.), *Harnessing human capital analytics for competitive advantage* (pp. 24–54). Hershey, PA: IGI Global.

Dahiya, R. & Rangnekar, S. (2018b). Forgiveness in Indian organizations: a revisit of the heartland forgiveness scale. *Current Psychology*, 1–18, doi: 10.1007/s12144-018-9879-6.

Dahiya, R. & Rangnekar, S. (2019a). Linking forgiveness at work and negative affect. *South Asian Journal of Human Resources Management*, *6*(2), 222–241.

Dahiya, R. & Rangnekar, S. (2019b). Relationship between forgiveness at work and positive affect: the role of age as a moderator. *International Journal of Environment, Workplace and Employment*, *5*(3), 247–268.

Dahiya, R. & Rangnekar, S. (2019c). Validation of the positive and negative affect schedule (PANAS) among employees in Indian manufacturing and service sector organizations. *Industrial and Commercial Training*, *51*(3), 184–194. doi: 10.1108/ICT-08-2018-0070.

Dahiya, R. & Rangnekar, S. (2020a). Does organisational sustainability policies affect environmental attitude of employees? The missing link of green work climate perceptions. *Business Strategy & Development*. DOI: 10.1002/bsd2.110

Dahiya, R. & Rangnekar, S. (2020b). Harnessing demographical differences in life satisfaction: Indian manufacturing sector. *International Journal of Business Excellence*. doi: 10.1504/IJBEX.2019.10023030 (in press).

Dahiya, R. & Rangnekar, S. (2020c). Relationship between forgiveness at work and life satisfaction: Indian manufacturing organizations. *International Journal of Business Excellence*. doi: 10.1504/IJBEX.2019.10020518 (in press).

Dahiya, R. & Rangnekar, S. (2020d). Validation of satisfaction with life scale in the Indian manufacturing sector. *Asia-Pacific Journal of Business Administration*, *12*(3/4), pp. 251–268. https://doi.org/10.1108/APJBA-03-2019-0045

Dutton, J. E., & Heaphy, E. D. (2003). The power of high–quality connections. *Positive Organizational Scholarship: Foundations of a New Discipline*, *3*, 263–278.

Fisher, C. D. (2010). Happiness at work. *International Journal of Management Reviews*, *12*(4), 384–412.

Fredrickson, B. L. (1998). What good are positive emotions? *Review of General Psychology*, *2*(3), 300–319.

Frenking, S. (2016). Feel good management as valuable tool to shape workplace culture and drive employee happiness. *Strategic HR Review*, *15*(1), 14–19.

Fridrickson, B. (2001). The role of positive emotion in positive psychology: the broaden and build theory of positive emotion. *American Psychologist*, *56*(3), 218–226.

Froh, J. J., Sefick, W. J., & Emmons, R. A. (2008). Counting blessings in early adolescents: an experimental study of gratitude and subjective well–being. *Journal of School Psychology*, *46*(2), 213–233.

Herrbach, O. (2006). A matter of feeling? The affective tone of organizational commitment and identification. *Journal of Organizational Behavior: The International Journal of Industrial, Occupational and Organizational Psychology and Behavior*, *27*(5), 629–643.

Hillebrandt, I., & Ivens, B. S. (2013). Scale development in employer branding. In *Impulse für die Markenpraxis und Markenforschung* (pp. 65–86). Wiesbaden: Springer Gabler.

Kuvaas, B. (2006). Performance appraisal satisfaction and employee outcomes: mediating and moderating roles of work motivation. *The International Journal of Human Resource Management*, *17*(3), 504–522.

Lilius, J. M., Worline, M. C., Maitlis, S., Kanov, J., Dutton, J. E., & Frost, P. (2008). The contours and consequences of compassion at work. *Journal of Organizational Behavior: The International Journal of Industrial, Occupational and Organizational Psychology and Behavior*, *29*(2), 193–218.

Milliman, J., Czaplewski, A. J., & Ferguson, J. (2003). Workplace spirituality and employee work attitudes: An exploratory empirical assessment. *Journal of Organizational Change Management*, *16*(4), 426–447.

Newstead, T., Dawkins, S., Macklin, R., & Martin, A. (2019). We don't need more leaders– we need more good leaders. Advancing a virtues–based approach to leader (ship) development. *The Leadership Quarterly*, doi:10.1016/j.leaqua.2019.101312.

Rego, A., Ribeiro, N., & Cunha, M. P. (2010). Perceptions of organizational virtuousness and happiness as predictors of organizational citizenship behaviors. *Journal of Business Ethics*, *93*(2), 215–235.

Salas-Vallina, A., & Alegre, J. (2018). Happiness at work: developing a shorter measure. *Journal of Management & Organization*, 1–21. https://doi.org/10.1017/jmo.2018.24

Salas–Vallina, A., Alegre, J., & Fernandez, R. (2017). Happiness at work and organisational citizenship behaviour: is organisational learning a missing link? *International Journal of Manpower*, *38*(3), 470–488.

Singh, S., David, R., & Mikkilineni, S. (2018). Organizational virtuousness and work engagement: mediating role of happiness in India. *Advances in Developing Human Resources*, *20*(1), 88–102.

Sobel, M. E. (1982). Asymptotic confidence intervals for indirect effects in structural equation models. In S. Leinhart (Ed.), *Sociological Methodology 1982* (pp. 290–312). San Francisco, CA: Jossey-Bass.

Spector, P. E., & Fox, S. (2002). An emotion–centered model of voluntary work behavior: some parallels between counterproductive work behavior and organizational citizenship behavior. *Human Resource Management Review*, *12*(2), 269–292.

Tsachouridi, I., & Nikandrou, I. (2019). The role of prosocial motives and social exchange in mediating the relationship between organizational virtuousness' perceptions and employee outcomes. *Journal of Business Ethics*, 1–17. https://doi.org/10.1007/s10551-018-04102-7

Weiss, H. M., & Cropanzano, R. (1996). Affective events theory: a theoretical discussion of the structure, causes and consequences of affective experiences at work. *Research in Organizational Behavior*, *18*, 1–74.

Wright, T. A., & Goodstein, J. (2007). Character is not "dead" in management research: a review of individual character and organizational–level virtue. *Journal of Management*, *33*(6), 928–958.

11 Techno Innovative Tools for Employer Branding in Industry 4.0

Ravindra Sharma, Geeta Rana, and Shivani Agarwal

INTRODUCTION

EMPLOYER BRANDING

The term "employer brand" came into light in the year 1990 to management audience by Simon Barrow (Chairman of People in Business) and Tim Ambler (Senior Fellow of London Business School) in the *Journal of Brand Management* in December 1996. According to Minchington (2005), employer brand act as an "image of an organisation as a place to work." They defined employer brand as the package of functional, economical, and psychological benefits provided to the employer and employing companies.

Employer branding is the employer reputation at the place he works. It is the process of managing or maintaining of reputation in the workplace or among the job seekers, employees, and stakeholders. The employer's reputation is the reputation of the organization as an employee or can be said that it highlights you and your personality as who you are. It is said not to be in tangible form but an employer brand is the thing that you carry itself with you. It is said to be an asset that an employee cultivates for himself. Employer branding in simple is how you present your organization in front of the job seekers. It is the process of showcasing the organization's unique cultural processes and their application in the organization.

An employer brand can be termed as a "talent" or "people" brand. It tells that the organization is a great place to work and enhance the recruitment process as well increases the engagement and retention of employees in the organization. It is the effort of the organization to retain and recruit the employees by confronting them the place is ideal to do work and is a desirable place to do job. It is an effective strategy to stand in the competitive market and brigs human resource into the organization as it works on the assumption that human capital brings value to the firm and skillful human resource can lead to attain competitive advantage in this huge competition. As employer brand is termed to be a "reputation of a place to work" and attracts the human resource to work in the organization and making them to know that it is the desirable place to work by attracting them through their working conditions and workplace as well as culture and its application. Sharma, Singh, and Rana (2019) viewed that employer branding helps in attracting best talent for the organization.

It is the duty of the organization to attract highly skilled workers and make them to do work in the organization by attracting them through employer branding. External market helps to attract highly skilled individuals for the organization and make them to work in it, whereas internal market helps the organization to create workforce by continuously exposing the employees toward the value of an organization's employer brand and its unique workplace, culture, and other prerequisites to attract them to do business in their own unique way. Internal market helps the employees to know that the organization is a great place to work, thus making the employees to retain in the organization.

Employer branding is associated with the positive image of an organization by its existing employees and those who wish to associate with the organization if an opportunity given. This image of the organization strengthens its overall value proposition.

REVIEW OF LITERATURE

Employer Branding

Clark and Payne (1997) stated that branding plays a link between the consumer and marketer. It is the name which a consumer values and provides preference. Sutherland, Torricelli and Karg (2002) were of the opinion that employer branding helps the employees to apply for the best jobs in the best organization as it provides knowledge about good organizations and make them to apply into it, and thus provide organizations with good talents without taking much efforts in finding them. Rana and Sharma (2019a) postulated that employer branding practices help in enhancing employee engagement in the organization. Fulmer, Gerhart and Scott (2003) found that employer branding provides stability and high positive workforce and had a positive effect on an organization's performance. Fernon (2008) stated that employer branding has the ability to retain employees by providing them the environment that allows them to live the brand through various aspects such as training and progression. Armstrong (2007) was of the opinion that employer branding aim is to become employer choice. It provides an organization to be able to attract fresh talent, recruit them, and to retain them in this growing shortage of labor in a competitive environment. Chhabra and Sharma (2014) postulated that companies should understand the importance of employer branding and use it as an instrument to allow the competitors to themselves as how they are different from them and to show their capabilities and capacities with good brand image in the market. Frook (2001) in his study found that internal branding helps the organizations to create a culture of trust between employers and employees. Burman and Zeplin (2005) and Martin, Gollan, and Grigg, (2011) found in their study that internal branding helps the organization to build a strong positioning of an organization that is difficult to beat by any competitors. Lievens et al. (2007) stated that employer branding is essential not only externally to attract fresh talent but also internally as it satisfies the employees and leads them to get motivated and be retained in the organization. Sharma, Singh, and Rana (2019) determined in their study conducted on Indian IT sector that organizational commitment can be sustained though employer branding practices. Maxwell and Knox (2009) studied the need of employer brand in an organization and in its recruitment process. Knox and Freeman (2006) postulated that employer branding consists of three important components, that is, internal branding, external branding, and potential talent in the job market. Their perception about the organization creates brand building for employer. It plays an important part of an organization as it leads to candidates to attract towards the organization by keeping in mind the employer brand of that organization and apply for that particular organization. Backhaus and Tikoo (2004) were of the opinion that if an employee is committed to the brand of an organization, he or she will not leave the organization even if he or she got an offer from other employers. Numerous researches have been conducted to determine the vitality of employer branding for organization and to sense the areas where organization have to focus to create an effective brand for the organization According to. Randstad (2018) a financially healthy organization is able to promise and deliver stability to the employees. Shitika, Tanwar, and Shrimali (2013) in their study identified 11 important items which are vital for developing an effective employer brand. These are company image, employee engagement, public relation, brand portfolio, employee satisfaction, working environment, internal recruitment, employment offerings, communication channel, and strategic policy. According to Wilden et al. (2010), employer branding incurs large amount of money on maintaining career websites, career fairs, internships, organizing competition on employer-of-choice award to develop their talented employees. Moroko and Uncles (2008) were of the view that in every stage of employment lifecycle employer branding plays a crucial role that cannot be denied by the organization at all. Still a lot of research is required to understand this concept of employer branding in more departments ((Davies, 2008)

Human Resource Management and Industry 4.0

Industry 4.0 is considered as an era of techno innovations and implementation; it stresses more on automation and digitalization. Some of the major components of industry 4.0 are artificial intelligence (AI), big data analytics, augmented reality, virtual reality, cloud computing, and internet off things. No domain in the field of business is untouched where these technologies have not proved its relevance. Human resource management is also one of such important areas where industry 4.0 have produced extra ordinary results and transformed entire human resource management practices. The changes performed in HR operations impacted the overall functioning of the organization; therefore, it is very much necessary that adoption of such technologies in HR function must be chosen with utmost care. In HR.4.0 employer branding in one of the important components which had impacted every stage of manpower handling from recruitment to superannuation. To pay attention in this area has become need of the hour for human resource management practitioners. The current industry 4.0 technology is effectively used by HR practitioners to manage next generation employees. (Hecklau et al., 2016). AI in Industry 4.0 is bringing forth these unexpected changes every day (Piwowar & Katarzyna, 2020). In industry 4.0 Cloud-based human capital management solutions, insight-based analytics, and democratized dashboards will enable the professionals to create learning, collaborative, and interactive talent scape for agile organization (Rana, G. & Sharma, R., 2019b).

The industry 4.0 has been highly focused on automation, and many manual jobs are at risk. There are very few jobs that are qualified to be resistant of the automation brought in by the industry 4.0 (Degryse, 2016). The strategic management of competencies of the workforce in order to bridge the gap between the industries requires knowledge and the skills of the employees (Hecklau et al., 2016). The notion of HR 4.0 is derived from the popular term "Industry 4.0," or "the fourth industrial revolution" (Vaidya et al., 2018).

OBJECTIVE OF THE STUDY

The objective of this research is to understand the employer branding and industry 4.0 – that is, how the innovative practices have evolved in industry 4.0 to promote employer brand for the organization. A conceptual model has been also suggested that depicts the relationship between employer branding and industry 4.0, its outcome in terms of new employer branding promotional activities.

CONCEPTUAL FRAMEWORK

Conceptual Model: Integration of Technology in HR Functions for Enhancing Employer Branding in Industry 4.0

The above framework depicts the relationship between HR components that organization wants to present project before the existing employees and potential candidates, components of industry 4.0 and its technological tools which develop through industry 4.0 components and helps in promoting HR components for organizational brand building (Figure 11.1).

Employer brand building is significant for organizational sustainable growth. Having strong employer brand means that the current as well as former employees

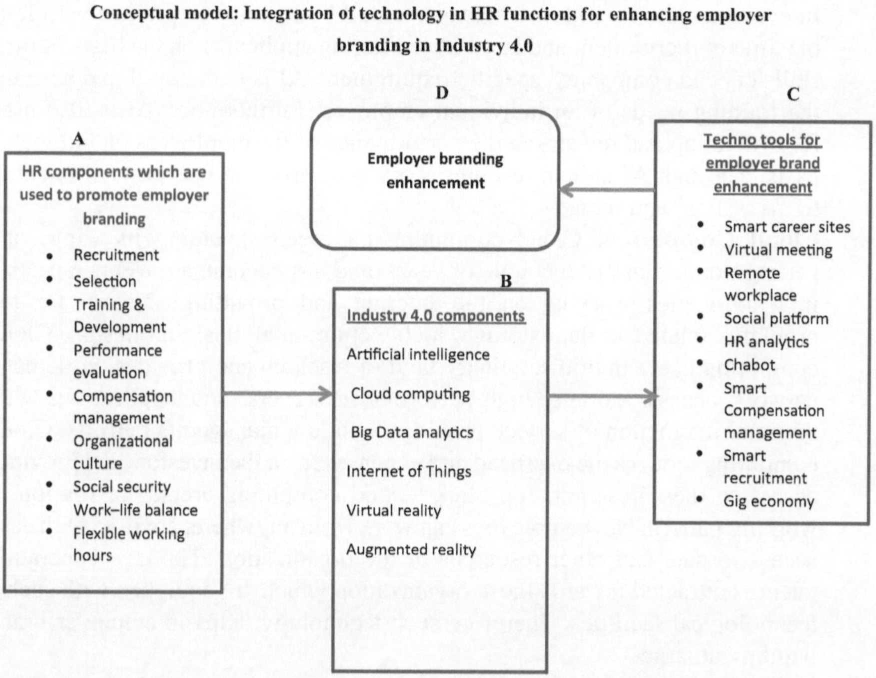

FIGURE 11.1 Integration of technology in HR functions for enhancing employer branding in Industry 4.0.

(Source: Authors' own work.)

are satisfied with the organization as an employer. Portraying organization as a great place to work is not an easy task. A number of disruptive technologies will continue to affect the employer branding efforts. These technologies are playing pivotal role in brand building and spread brand message among employees, potential candidates, and other stakeholders. Here we are discussing about few of the techno innovative tools that organizations are adopting for making employer brand building in Industry 4.0 era. Therefore, it's time to leverage these trends to brand's advantage. Here are some disruptive technologies that affect employer branding. The organization can include these technologies as a part of their employer's branding and recruitment marketing efforts. And then devise strategic objectives to fulfil the organization's short- and long-term goals.

a) **Artificial Intelligence and Employer Branding Building**: Adoption of new technology not only enhanced the overall business performance but also attracts and ensures sustainability among the young talent (Singh et al., 2019). AI is one of the important elements of industry 4.0 which is wholeheartedly adopted by human resource practitioners. Innovations that stem from AI include chat bots, which utilize machine learning algorithms to help in augmenting employer branding. Companies have found that these chat bots aid in strengthening their customer service and communication with potential candidates prior to their applications.

In every stage of workforce management, AI plays a very productive role at the time of recruitment and is used for filtering applicants on the basis of their skill sets and companies' specific requirement. AI is being used to determine the training needs of an individual employee. Furthermore, AI is also used for evaluating and rewarding the performance of the employees on individual basis. Through AI human resource work gets easy and it is quite efficient in terms of time and money.

b) **Cloud Computing:** Cloud computing has been spreading its wings and getting popular in just a couple of years, and has become a novel perspective in technologies working on the Internet and providing solution for the problems related to data storage, architecture, analytics, and design. Cloud computing is the multidisciplinary field of research and provides on-demand network access, storage, high-performance servers, and applications with minimal interaction of service providers and low management efforts. Cloud computing reduces the overhead and expenses from the investors and provides access to these resources on click. Cloud computing creates revolution in working pattern. Now employees can work from anywhere, anytime with easy access to data and other resources of the organization. The new generation talent is attracted towards those organization which are equipped with such a technological facilities. Therefore such technology helps in employer brand building at large.

AR and VR: One of the advanced technologies brought by the Industry 4.0 is the Augmented Reality (AR) and Virtual Reality (VR). AR actually adds digital elements to the physical spaces, whereas VR creates a simulation of the virtual world with digital elements, completely shutting the user from the physical world. From the game environment, both AR and VR have invaded into the space of human resource departments inside the organizations. In recent years, AR and VR devices have been developed for applications in the workplaces. AR and VR have started to find place in larger organizations to provide a constant learning environment for the employees, and are useful in simulating the actual work environment which avoids searching through complex online sources or outdated printed manuals.

c) **Compensation Management Tools:** Nowadays companies are using specialized high tech tools of compensation management and ensuring the compensation schemes which are performance based as well as decrease gap and dissatisfaction related to compensation. Technology is taking over most of the compensation related part in organizations (Othman et al., 2017). Payroll processing, pay fixation according to employee experience and performance, customized salary packages are being built and managed through technology, which indirectly increases employee trust toward the organization which leads to better employer branding. In the current era, the components of compensation are managed through technology only.

d) **HR Analytics:** People Analytics sometimes can be referred to as talent analytics or HR analytics. In this regard, it may be deeply defined as the data-driven and goal-focused method to study the process related to people, their functions in organization, challenges, and the opportunities they face at work in order to elevate these systems and help people and organizations achieve business success. This involves using the data to take better decisions like hiring, performance appraisal, and increasing the workforce in an organization and other actions pertaining to the workforce with great accuracy and precision. HR analytics may also be seen as the methodological identification and quantification of employee related drivers and metrics that ultimately help the organizations attain the business objectives (Heuvel & Bondarouk, 2016).

e) **Glass Door:** Glass door web resource is a very popular platform where companies can introduce their virtual office for the job aspirants so that they can understand the working culture before joining the organization.

f) **LinkedIn:** LinkedIn is a social network for the business community. LinkedIn consumers generate specialized, review-similar outlines that permit other location associates to acquired additional around their commercial experience, their expanses of proficiency, and collections or administrations they are appropriate toward. Once users create their profile, they can add other operators to their system. LinkedIn is a social network that focuses on professional networking and career development. Many companies are using LinkedIn to promote them as a best employer in the industry.

g) **Instagram and Twitter:** Instagram and Twitter are organized communal spreading positions measured for circulation distribution and fulfilled. Twitter features retweeting, quoting, and multilevel response chains, while Instagram has only single degree reply chain. Starbucks does an excellent task in making sure that they domesticate a strong community among their employees. For instance, they talk to contemporary employees as partners, instilling a sense of pleasure in every employee. Moreover, Starbucks produced Instagram and Twitter responsibilities particularly for @Starbucks Occupations, which they use to encourage their innovativeness product and involve with development investigators.

h) **Smart and Interactive Career Sites:** With the help of advance technology integration, organizations are coming up with career sites that are more candidate-friendly and cable of resolving aspirants' queries on real time basis. Career site of a company is one of the prominent places where the company can demonstrate itself as the best place for working. Nowadays companies are using career site for employer branding promotion, which yields very positive results.

i) **Banners:** Banners are used as a good source of employer branding as it is used to attract employees through Online Banners or Street Banners. Online Banners provides the facility of flashing facility of different web pages of applicant choice and price.

j) **Social Networking Platforms:** If any company wants to reach masses, social networking platform is one of the most effective and rapid methods. The old phrase "The network is your net worth" becomes more relevant with the inclusion of social networks in company branding strategies. The wise use of these platforms helps in employer branding building of the organization.

k) **Job Portals:** Earlier job portals were used only for job posting by the organizations. But now these job portals become a hot platform for companies to demonstrate their employer branding through landing page, talent network forums, connection with future applicants, sharing the company culture and vision, and employees' welfare initiative through this platform has become prominent. The site is also an intelligent way to share approaches, instances, and the philosophy driving your employer branding initiative, as well as the company's aspirations for its employees, and its overarching goals.

l) **Strong Email Marketing Campaign:** Often, locating the right fit talent for a role is all about timing and personalization. Any recruiter who's eager to excel in their role will build a targeted, bespoke, and innovative email campaign – detailing the position, success stories, quotes, and organizational culture, evoking interest and enthusiasm for the role. Again, this is a must-have for any employer branding blueprint. To really help the messaging standout, look at a tool like JotForm – the solution can draw data from online forms, adding customized elements to PDFs. This means employers can tailor signatures, greetings, and design elements, when disseminating marketing emails, in bulk, with little effort.

m) **Gig Economy – Remote Work:** The concept of a gig economy involves project-based workers, freelancers, temps, and independent contractors. The Gig Economy Data Hub says that this offers companies with greater flexibility and leeway to outsource work when necessary. Gig economy positions ensure greater economic security for aspiring candidates, especially for those who are struggling to find more permanent positions. The idea of remote work is similar to this. It appeals to candidates who find creativity and greater productivity outside of a fixed office environment. There is a number of software such as video meetings and other communication platforms which helps in seamless communication. It also helps in team collaboration and makes all members feel accountable for their work. In addition to full-time employees, the prospect of remote work in companies shows that companies put a premium on the work–life balance of employees. It is found that 80% of employees are more likely to pick a job with flexible work options. It is important to remember that remote work is only possible when the entire process is in place. You ought to take care to see that it fosters consistency among the team members regardless of where they work.

CONCLUSION

Companies can gain competitive advantage if they succeed to create a brand image which leads to talent attraction and retention. The role of technology is vital in creating the brand image of the firms. Technology tools spread the employee branding-related initiative of the firms among internal employees and prospective candidates. Therefore, we can conclude that organizations should use technological tools intensively but wisely to promote themselves as the best employer in the job market.

REFERENCES

Armstrong, M. (2007). *Employee Reward Management and Practice*. London: Kogan Page Limited.

Backhaus, K., & Tikoo, S. (2004). Conceptualizing and researching employer branding. *Career Development International*, 9, 501–517.

Burmann, C., & Zeplin, S. (2005). Building brand commitment: A behavioural approach to internal brand management. *Journal of Brand Management*, 12(4), 279–300. doi: 10.1057/palgrave.bm.2540223.

Chhabra, N. L., & Sharma, S. (2014). Employer branding: Strategy for improving employer attractiveness *International Journal of Organisational Analysis*, 22(1), 48–60.

Clark, M. C., & Payne, R. L. (1997). The nature and structure of workers' trust in management. *Journal of Organizational Behaviour*, 18, 205–224. doi:10.1002/(SICI)1099–1379(199705)18:3<205::AID-JOB792>3.0.CO;2-V.

Davies, G. (2008). Employer branding and its influence on managers. *European Journal of Marketing*, 42(5/6), 667–681.

Degryse, C. (2016). Digitalization of the economy and its impact on labour markets. *ETUI Research Paper-Working Paper*.

Lievens, F., Van Hoye, G. and Anseel, F. (2007), Organizational identity and employer image: towards a unifying framework*. *British Journal of Management*, 18, S45–S59. https://doi.org/10.1111/j.1467-8551.2007.00525.x

Fernon, D. (2008). Maximising the power of the employer brand. *Admap*, 494, 49–53.

Frook, J. E. (2001). Burnish your brand from the inside. *B to B*, 86, 1–2.

Fulmer, I. S., Gerhart, B., & Scott, K. S. (2003), Are the 100 best better? An empirical investigation of the relationship between being a "great place to work" and firm performance. *Personnel Psychology*, 56, 965–993. doi:10.1111/j.1744–6570.2003.tb00246.x.

Hecklau, F., Galeitzke, M., Flachs, S., & Kohl, H. (2016). Holistic approach for human resource management in Industry 4.0. *Procedia Cirp*, 54(1), 1–6.

Knox S. & Freeman C. (2006). Measuring and managing employer brand image in the service industry. *Journal of Marketing Management*, 22(7–8), 695–716. DOI: 10.1362/026725706778612103

Martin, G., Gollan, P. J., & Grigg, K. (2011). Is there a bigger and better future for employer branding? Facing up to innovation, corporate reputations and wicked problems in SHRM. *International Journal of Human Resource Management,* 22(17), 3618–3637. doi: 10.1080/09585192.2011.560880.

Maxwell, R., & Knox, S. (2009). Motivating employees to "live the brand": A comparative case study of employer brand attractive-ness within the firm. *Journal of Marketing Management*, 25(9), 893–907.

Minchington, B. (2005). *Your employer brand—Attract, engage, retain*. Torrensville: Collective Learning Australia.

Moroko, L., & Uncles, M. D. (2008). Characteristics of successful employer brands. *Brand Management*, 16(3), 160–175.

Othman A. K., Hamzah M. I., Abas M. K., and Zakuan N. M. (2017). The influence of leadership styles on employee engagement: The moderating effect of communication styles. *International Journal of Advanced and Applied Sciences*, 4(3), 107–116.

Piwowar-Sulej, K. (2020). Human resource management in the context of Industry 4.0. *Organization and Management,* 1(49), 103–113.

Rana, G. & Sharma, R. (2019a). Assessing impact of employer branding on job engagement: A study of banking sector. *Emerging Economy Studies*, 5(1), 7–21.

Rana, G., & Sharma, R. (2019b). Emerging human resource management practices in Industry 4.0. *Strategic HR Review*, 18(4), 176–181. https://doi.org/10.1108/SHR-01-2019-0003

Randstad (2018). *Employer Brand Search Global Report*.

Schneider, L. (2003). What is branding and how is it important to your marketing strategy? Available at: http://marketing.about.com/cs/brandmktg/a/whatisbranding.htm (accessed 22 October 2003).

Sharma, R., Singh, S. P., & Rana, G. (2019). Employer branding analytics and retention strategies for sustainable growth of organizations. In: Chahal, H., Jyoti, J., & Wirtz, J. (Eds.), *Understanding the Role of Business Analytics*. Singapore: Springer. https://doi.org/10.1007/978-981-13-1334-9_10.

Sharma, R., Jain, V., & Singh, S. P. (2018). The impact of employer branding on organizational commitment in Indian IT sector. *IOSR Journal of Business and Management*, 20(1), 49–54. https://doi.org/10.9790/487X-2001054954.

Shitika, T. S., & Shrimali, V. (2013). Modelling effectiveness of employer branding - An interpretive structural modelling technique. *Pacific Business Review International*, 5(11), 82–89.

Singh R., Anita G., Capoor S., Rana G., Sharma R., & Agarwal S. (2019). Internet of Things enabled robot based smart room automation and localization system. In: Balas, V., Solanki, V., Kumar, R., & Khari, M. (Eds.), *Internet of Things and Big Data Analytics for Smart Generation*. Intelligent Systems Reference Library, vol. 154. Cham: Springer. https://doi.org/10.1007/978-3-030-04203-5_6.

Sutherland, M. M., Torricelli, D. G., & Karg, R. F. (2002). Employer-of-choice branding for knowledge workers. *South African Journal of Business Management*, 33(4), 13.

Vaidya, S., Ambad, P., & Bhosle, S. (2018). Industry 4.0–a glimpse. *Procedia Manufacturing*, 20, 233–238.

Wilden, R. M., Gudergan, S., & Lings, I. N. (2010). Employer branding: Strategic implications for staff recruitment. *Journal of Marketing Management*, 26(1–2), 56–73.

12 Impact of Employer Branding on Customer Acquisitions and Retentions

A Case Study of Microsign Products

Ramzan Sama

INTRODUCTION

The inclusion of specially-abled in the workforce is a core of corporate social responsibility (Kuznetsova, 2012). It is labeled as corporate social responsibility (CSR) by media houses (Bhatia, 2008; Bhattacharjee, 2011). Specially-abled people have been marginalized, in India due to the perception of "non-employable" (Sama, 2020). Small and medium enterprises (SMEs) employing 40% of the workforce, thus boosting the Indian economy (Goyal, 2013). Microsign, an SME, has 60% specially-abled workforce. The basic focus of employer branding so far was on external branding and less on internal employee satisfaction. Past research has established the link between employer brand and corporate brand management (Mosley, 2007). Organizations emphasized more on human resource to build corporate image as a good employer brand leads to good corporate image among the stakeholders (Weiwei, 2007). There is limited research on exploring the impact of corporate image on consumer brand relationship (Trivedi, 2020). Further, employees can be a good source for creating a corporate brand (Keller, 2000). However, the impact of employer branding on customer retention and acquisition in SMEs is unexplored (Wilden et al., 2010). Thus, it is interesting to study the employment of specially-abled in SMEs,

Hence, this study in the form of a case study of Microsign Products (an SME employing 60% specially-abled employees) examines its policy of hiring specially-abled people and CSR and its impact on employer brand. Further, the impact of Microsign employer brand on corporate image and corporate brand and the outcomes, namely, customer acquisition and retention are examined.

EMPLOYER BRANDING

Brand management core concern is to deliver a consistent and distinct brand image. Aaker (1993) pioneered the brand management concept, postulate that brands are a powerhouse in building a relationship in "producer–seller–investor–buyer (consumer)" system. Ambler and Barrow (1996) coined the term "Employer Brand." They defined it as "the package of functional, economic and psychological benefits provided by employment and identified with the employing company." The rationale behind the employer brand was to give the same emphasis on organizations employee proposition as customer brand propositions. The basic focus of employer branding so far was on external branding and less on internal employees' satisfaction. Barrow and Mosley (2005) propounded employer branding as an integrated approach of recruitment, employee engagement, and customer brand equity. Mosley (2007) explored the link between employer brand and customer brand management.

Keller (2000) posited that the employees can be a good source for positioning and strengthening the corporate brand. Knox and Freeman (2006) argued that marketing and HR should work in an integrated way. The employer branding for internal environment is a narrow approach as customers also take the note of good HR practices. The positive HRM leads to publicity resulting into the attractive brand among the customers (Sharma et al., 2019). However, there is limited research on how employer brand influences customer perception and intention to buy (Anselmsson,

Bondesson & Melin, 2016). Furthermore, there is limited literature on the link between the employer brand and its subsequent impact on corporate brand and customer acquisition and retentions.

CORPORATE IMAGE AND CORPORATE BRAND

Corporate image cites the firm's reputation thus it is an asset for the firm. The corporate image is also the perceived image of the public when they see the company name. Rana and Kapoor (2016) posited that the corporate image associations include corporate name, culture, employees, ethics, customers, products, clients, shareholders. The marketers employ activities for cultivating the good corporate brand image in the market. However, extant research on the corporate image doesn't offer the universally accepted definition (Kennedy, 1977; Hatch & Schultz, 2003; Balmer, 2001; Gioia et al., 2000). Organizations use many integrated marketing communications tools like advertising, promotion and publicity to build the corporate image. But they focused more on human resources to build a corporate image as it is believed that good employer brand leads to good corporate image in the market. (Weiwei, 2007)

Keller (2000) posited that one of the pillars of building a corporate image and translating it to brand equity is the employee in the company. This strategy conveys the importance of employees to its customers. A study by Anselmsson et al. (2016) established the relationship between employer brand and customers' willingness to buy the product from the branded retail stores and found the empirical data in support of this link.

Andreev (2013) defined the term corporate brand as "a set of visual and verbal elements of company's brand that transmit its competitive advantages to target groups: employees, investors, distributors and consumers, society and the state." De Chernatony (1999) explained the use of corporate branding in marketing and the importance of people in building relationships with customers. The fine-tuning of HR with the corporate brand can be achieved by communicating the unique characteristics of employment to the target audience (Mosley, 2007).

Aaker (2004) argued that companies creating a corporate brand through concerned about the social and environmental cause are respected and liked more in the market resulting in positive attitudes and loyalty even in challenging times. Thus, the corporate brand can be a differentiation point for the organizations in a cluttered market.

The corporate brand carries heritage and plays a key role in a customer relationship (Aaker, 2004). This study is about the heritage of Microsign in hiring specially-abled people since1986, which was far before the UN disability act, 1993 and India's Persons with Disabilities Act, 1995 (Sama, 2020). This study aims to unravel the impact of employer brand on corporate image and corporate brand and its subsequent effect on customer acquisitions and retention of Microsign. In the academic literature, past research is limited to exploring this connection in the context of SMEs.

SPECIALLY-ABLED

The United Nations Sustainable Development Goals (SDGs) give guidelines to the international community for the specially-abled inclusive development. The specially-abled and their families faced more economic and social disadvantage compared to their normal counterparts. The unemployment rate is also higher among specially-abled compared to their normal counterparts (U N Flagship Report on Disability and Development, 2018).

Ruhindwa et al. (2016) posited that specially-abled are not considered as an abled employee and thus face exclusion from the mainstream employments. However, specially-abled are willing to earn to get financial freedom but they face difficulties in securing permanent employment (Bruce, 2006; Waterhouse et al. 2010). The inclusion of specially-abled in the workforce is key to social inclusion and also to improve their lifestyle (Yeend, 2012; Commonwealth Department of Social Services, 2011).

Extant research attributed that not their impairments but the organizational, social, attitudinal barriers (Morre & Fishlock, 2006), low education levels and social skills (Thomas & Lahla, 2002) are responsible for the exclusion of specially-abled people from the mainstream employment. In contrast, past researchers have found that specially-abled are equal or in some case more productive than their abled counterparts (Graffam, Smith, Shinkfield, & Polzin, 2002; Lewis & Priday, 2008).

According to WHO (2018), 15% of the world's population are facing some form of impairments. The United Nations Convention on the Rights of Persons with Disabilities (CRPD) promotes the full integration of persons with disabilities in society.

In India, 2.21% of the population are specially-abled (Statista, 2011). Only 26% of them get jobs. For specially-abled inclusion, various laws have been enacted like Mental Health Act 1987 and the Persons with Disabilities (Equal Opportunities, Protection of Rights and Full Participation) Act 1995.

Specially-abled have been marginalized in India due to the perception of "non-employable." SMEs employ40% of the workforce, thus boosting the Indian economy. India has the second largest SMEs (48 million) in the world after China with 50 million (Goyal, 2013). Thus, it is interesting to explore the case of SMEs in terms of offering employment to the specially-abled people and its impact on employer branding leads to successful customer brand management. Siperstein et al. (2006) studied the consumers' attitudes toward the firm hiring specially-abled people and found that consumers have strong beliefs in the companies hiring specially-abled and they give preference to such companies in giving them business. This study aims at exploring this further by mapping the impact of employer branding and corporate image due to the employment of specially-abled and its impact on the consumer-brand relationship, in case of SMEs with special reference to Microsign.

MICROSIGN PRODUCTS

Microsign Products (Microsign) founded in 1978 by Nisheeth Mehta, in the small-town Bhavnagar in Gujarat. The company produces the components for electronic, automobiles and defense industries. Their main products are viz. plastic fasteners, clips, and clamp closers. Table 12.1 shows a list of the products. Their clients are big companies and MNCs like Tata, Honda, Maruti, ISRO, DRDO, and Reliance

TABLE 12.1
List of Products

Product segments: Electronics-Auto-Defense-Aerospace-Packaging

Standard products	*Tailor-made products*
Cable tie releasable double-headed and beaded type (multi-purpose)	Cable tie-facility type
Cable tie non-releasable (self-locking)	Zipper (quick-clip)
Cable tie non-releasable (self-locking)	Cable clamps and cord clips (adhesive-backed)
Tie mounts (adhesive-backed)	Rivet & Furr clips
Various kinds of cable identification marking system	Closure tree clips & Fastener

Source: Sama (2020). Envisioning a mutually inclusive growth story: A case study of Microsign Products.

TABLE 12.2
List of Award Received by Microsign

1	National Award for Welfare of People with Disabilities (1999) from Government of India, Ministry of Social Justice & Empowerment. The Award was handed over by then President of India- Shri. K R Narayanan.
2	NCPDP Helen Keller Award (1999) by Rajiv Gandhi Foundation, for promotion of employment of disabled people.
3	FICCI Annual Award (2003-2004) in recognition of corporate initiatives in the empowerment of Physically Challenged.
4	IMC Ramakrishna Baja National Quality Award (2007) Special Award for Performance Excellence, in small business category.
5	CNN-IBN Award of "Real Heroes of India" (2009) for giving employment to differently-abled people.

Source: Sama (2020). Envisioning a mutually inclusive growth story: A case study of Microsign Products.

to name a few (Table 12.3). Products made by the Microsign are quality products certified by the ISO/TS 16949 and IATF 16949 third party quality certifications for the automobile industry. Today, Microsign with 60% of specially-abled workforce supplying quality products to all its multinational clients and breaking all the success and profits of the business.

According to Nisheeth, they never discriminate between abled-bodies and specially-abled employees; appraisal is entirely based on performance. Due to these noble works, the company bagged more than 10 reendowed awards including NCPDP Helen Keller Award (1999) and National Award for Welfare of People with Disabilities (1999) from Government of India.

Microsign mission in words of Nisheeth "we want to make specially-abled people self-sustainable and live dignified life by optimum utilization of their strengths, by creating a win–win situation for organization and specially-abled employees."

TABLE 12.3
List of Clients

PSUs	Indian Private	MNCs	Defense and Space
Bharat Heavy Electricals Ltd.	TATA Motors Ltd.	Siemens AG	Defense Research and Development Organisation (DRDO)
Bharat Dynamics Ltd.	Larsen and Toubro Ltd.	Honda Motor Co. Ltd.	Indian Space Research Organisation (ISRO)
Hindustan Aeronautics Ltd.	Sun Pharmaceutical Industries Ltd.	Volvo Car Corporation	
Bharat Electronics Ltd.	Hero Motocorp Ltd.	Hewlett Packard Company	

Source: Sama (2020). Envisioning a mutually inclusive growth story: A case study of Microsign Products.

MICROSIGN HISTORY

Nisheeth after pursuing civil engineering wanted to pursue his passion for manufacturing plastic products. He started small plastic molds manufacturing company in the name of Microsign Products, popularly known as Microsign. His startup did well in the initial years. In the expansion phase, when he was searching for employees, he thought of hiring specially-abled people to include them in mainstream business. When asked what drives him to do this social experiment in the growth stage Nisheeth said, "My niece, has a hearing impairment and despite education in the top school of Chennai, she faced social discrimination in getting employment due to the impairments." This incidence in Nisheeth own family changed his thinking toward the specially-abled people.

While Nisheeth started hiring specially-abled people, he observed that specially-abled people were not getting an opportunity because of various reasons like social barriers, transportation issues, requirements of special infrastructure at firms and abled peers' attitude toward specially-abled employees. Despite all these challenges, Nisheeth listens to his inner voice and initially hired hearing and speech impaired persons as he knew sign language due to interactions with his niece. He faced many challenges in training and making them accepted among abled employees but after some time he was stunned by looking to the dedication and diligent work by these employees. Later on, Microsign also started hiring orthopedically challenged and mentally challenged people.

Nisheeth wanted to transform specially-abled abilities into their strength to give them dignified and respectful status in society. So, he adopted professional HR practices to tap their optimum potentials. He started implementing competency mapping of all the abled and specially-abled employees. When asked, how did he do competency mapping of specially-abled employees? He said that they put hearing-impaired employees on noisy machines, where an abled employee gets distracted due to noise but since hearing impaired employee could not hear, noise is not the distraction for him leading to the increased in productivity compared to an abled employee. Similarly, abled employees got bored in an assembly line by doing the repeated work. But in case of mentally challenged, once they learn the steps, they keep on doing the same steps without getting bored leading to increased productivity.

Hardik Shah, a mentally challenged employee of Microsign express the contribution of Microsign in his own words as" I got the status and confidence in myself after getting employment at Microsign. Earlier I was not able to do even my routine work also. Employment boost up my confidence and give me respect in society. Hardik and his mom got a chance to get appeared in the popular TV program on Microsign hosted by Mr. Amitabh Bachchan, Indian Bollywood star in "Aj Ki Raat Hai Jindagi" on Star TV channel. This was something a mentally challenged son, and his mother could never even dream of.

MICROSIGN OTHER CSR ACTIVITIES

The Centre for Excellence is a CSR wing of Microsign. This concept of excellence is based on the universal values of humanism. The center pursues conceptual and executional excellence in the realm of youth development, healthcare, education, women and child development, old age care, arts and culture.

The inclusion of specially-abled in the workforce is a core of CSR (Kuznetsova, 2012). There is limited research on the social inclusion of specially-abled people as a CSR (Bennett, 2011). Also, despite growing interest in CSR, little literature is available on hiring specially-abled as a CSR (Wang et al., 2016). Hiring specially-abled is core in Microsign CSR and it's in their heritage. The fair employee practices help organizations in developing social face in society (Matuska & Imińska, 2014). Mokina (2014) posits that there is a dearth of literature on employer branding as a part of corporate branding.

LITERATURE REVIEW

SPECIALLY-ABLED AND CSR

Griffin, Peters and Smith (2007) posited that specially-abled people either have to make them efficient as abled or have to stay away from the abled bodies. Keeping the disabilities aside and keeping humanitarian ground employers focus on their special abilities. However, the employment rate of specially-abled is much lower in the population (Bhalla, 2015; Mitra & Sambamoorthi, 2006). These numbers are even lesser in the case of the private sector (DEOC, 2009; NCPEDP, 1999; Somvanshi, 2015). Those who are working also get the stereotypical job of the clerk, machine operators, etc. (Bhadra, 2010). To overcome these challenges, employment of specially-abled needs to be a part of company's CSR activities (Monachino & Moreiram, 2014; Kuznetsova, 2012).

The inclusion of specially-abled is label as CSR by media houses (Bhatia, 2008; Bhattacharjee, 2011). These reports posit that inclusion of specially-abled does not put burdens on private companies but creates win–win situations for employee and employers (Ghosh, 2011; Kela, 2012; Rangan, 2011). Siperstein et al. (2006) found the positive relationship between consumer attitude and purchase intentions from the companies hiring the specially-abled persons.

Matuska and Sałek (2014) concluded that a diversified workforce has an impact on corporate identification and CSR both. Further CSR is a powerful tool of PR and building the brand in the external environment. Thus, there is a link between diversified workforce, employer branding and CSR.

Hidegh and Csillag (2013) posit that there are still physical and mental barriers in hiring the specially-abled people. As a result, specially-abled have lesser career opportunities compared to able-bodies (Berthoud, 2008). The employment of specially-abled as a CSR strategy is more visible in a large corporation and MNCs (Miethlich & Slahor, 2018). The size and resources of the company affect the employment policies (Bruyere et al., 2006). However, "specially-abled champions" companies are missing that could serve as a role model for other companies (Fasciglione, 2015; Hernandez et al., 2008). Inclusion of specially-abled persons in the workforce is a good way for CSR activities in the internal and external environment (Csillag & Gyori, 2016). However, the employment of specially-abled as CSR and evaluating best practices is understudied in academic literature (Perez et al., 2018; Miethlich & Slahor, 2018). This study examines the impact of hiring specially-abled on employer branding.

Employer Branding

Minchington (2005) defines employer brand as "the image of your organisation as a 'great place to work' in the mind of current employees and key stakeholders in the external market." According to Kimpakorn and Tocquer (2009), employer brand is how potential employees perceived a company. Lloyd (2002), Backhaus and Tikoo (2004) and Rana and Sharma (2019) emphasized that the engagement of existing employees should also be the part of employer. The antecedent to an attractive employer brand is an organizational image, as the positive image helps in attracting new employees and increase employee loyalty (Herrbach & Mignonac, 2004). Ambler and Barrow (1996) suggested the measurement of employer branding on three parameters' economic, functional and psychological benefits. Extant research proved the link between the employer's image and attraction and retention of employees (Cable & Graham, 2000; Greening & Turban, 2000). The past literature has also confirmed the relation between employer brand and job satisfaction and employee satisfaction (Helm, 2013; Priyadarshi, 2011; Du Preez & Bendixen, 2015; Herrbach & Mignonac, 2004). Employer branding is not only used for making an image of "employer of the choice" but it is also has been used as a tool for engaging customers (Martin et al., (2005; Barrow & Mosley, 2005). Potential employees are drawn to organizations that are identified as a good brand (Rana & Sharma, 2018).

Backhaus and Tikoo (2004) suggested that a competitive advantage can be achieved by employer branding. Gaddam (2008) emphasized that employer branding leads to employee satisfaction and commitment resulting in customer satisfaction and loyalty. Wilden et al. (2010) established the link between low brand awareness of a company among its consumers and awareness among the prospect workforce. Thus, employer brand does have many marketing implications. Moroko and Uncles (2008) argued that the customer brand aspects are important for employer brand and both are integrated and has an impact on the internal and external environment.

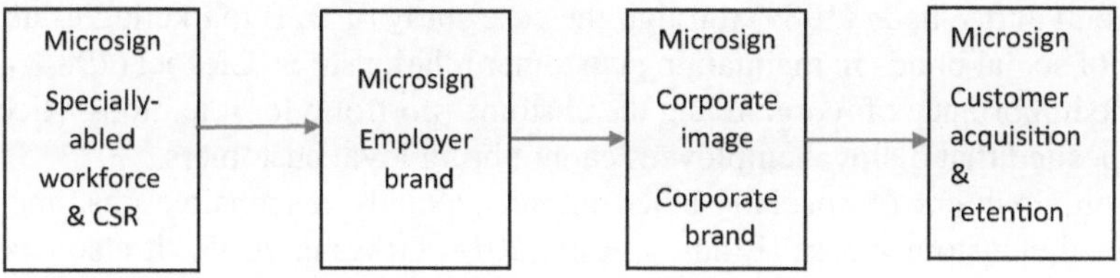

FIGURE 12.1 Author's proposed model.

CORPORATE IMAGE AND CORPORATE BRAND

The hiring of specially-abled creates a good corporate image and promotes the reputation of a socially responsible company (Miethlich & Oldenburg, 2019). The positive corporate image helps the organization in attracting employees (Fombrun & Shanley, 1990) and increase customer satisfaction and loyalty (Andreassen & Lindestad, 1998). It gives a distinct image in the minds of stakeholders (Kuznetsova, 2012; Gröschl, 2005). At the same time, it is also important to communicate the company's commitment to social inclusion to the outside world (Miethlich & Slahor, 2018). This leads to a stronger corporate brand (Csillag & Gyori, 2016). Work values of the organization help in building the brand image of the organization (Rana, Rastogi & Garg 2016).

Aaker (2004) posits that corporate brand heritage or roots help in struggling time. Aaker also argued in the same study that people play a key role in creating corporate brand leading to liking, respect and customer loyalty. Thus, citizenship corporate brands get benefits of doubts because of CSR activities. The past research on corporate branding posits the importance of workforce environment on corporate reputation (Fombrun & Gardberg, 2000). Keller (1993) and Kapferer (2004) postulated that to differentiate the brand in the market, corporate and consumer brand health is the key step in the brand-building process. The extant researches on corporate branding also argued that brand development process is in the purview of all stakeholders for aligning the values with employees and customers (Hatch & Schultz, 1997; 2001; Urde, 2013). Aaker (2003) posits the importance of focusing on the company's culture, values, and workers to create a corporate brand. Chernatony (1999) attributed the responsibility of building a corporate image to the HRM department. Tamberg and Badyin (2009) postulate the employer brand as a subset of the corporate brand. Mokina (2014) propounds the components of the corporate brand as a social brand, goodwill and employer brand. However, the research on the implications of employer brand on corporate branding is limited in academics research (Mokina, 2014). This research explores the impact of employer brand on a corporate image leading to a stronger customer-brand relationship.

CUSTOMER ACQUISITIONS AND RETENTIONS

The importance of customer retention was first explored by Dawkins and Reichheld (1990) who established the positive impact of customer retentions on customer net present value.

Turnbull and Wilson (1989) through the case study of B2B marketing established the role of social bonds in maintaining customer relationships. Ciezpel (1984) argued about the importance of reorganizing the customer portfolio for retentions. Riechheld (1996) posited that disloyal employees cannot bring loyal customers.

The employment of specially-abled attracts socially responsible customers and increases the customer base (Lindsay et al., 2018; Gröschl, 2005). It also ensures a good local client base (Dibben et al., 2002). This practice of hiring specially-abled also improve the image and increase the customer satisfaction resulting into retention (Lindsay et al., 2018) and improve overall customer relations (Csillag & Gyori, 2016; Kuznetsova, 2012). It is also found such type of organizational justice practices enhances the managerial effectiveness (Ranaand Rastogi, 2015)

Several research has studied the impact of hiring specially-abled on customers acquisitions and retentions in general (Lindsey et al., 2018; Suter, 2007). However, the impact of employer branding on customer retention and acquisition in SMEs is unexplored (Wilden et al., 2010). This research aims to fill this gap in the literature.

MICROSIGN SPECIALLY-ABLED WORKFORCE AND CSR

The hiring of specially-abled people is in the heritage of Microsign. They started hiring specially-abled way back in 1986 when there were no reservations or act for specially-abled in private or government organizations. To understand the working of specially-abled employees, some of them were interviewed.

Tukesh Bhatt (60) a deaf and mute person was the first employee of Microsign in 1986. Before Microsign, he was not getting entry in the mainstream business because of impairments. Nisheeth gave him a chance to work in a mainstream company. Tukesh retired after working for 32 years at Microsign. Today, he is living a happy life with his wife and two children. He feels proud to retire as a mainstream worker and it boosts his confidence.

Another employee, Hardik Shah, is a mentally challenged working in a packaging department of Microsign. His mother Meena said, "Earlier Hardik was not able to do the routine chores but after getting employment and constant care and love from Nisheeth, now Hardik is a confident person. She and Hardik got a chance to get appeared in the popular TV program on Microsign hosted by Mr. Amitabh Bachchan, Indian Bollywood star in 'Aj Ki Raat Hai Jindagi' on Star TV channel." This was something a mentally challenged son, and his mother could never even dream of.

Today, Microsign with the 60% specially-abled workforce breaking all the heights of success in the business. Nisheeth said that the belief in converting the disability into abilities by doing perfect competency matching of specially-abled persons through various job rotation. For example, a deaf person placed on injection machinery does not get distracted by Noise and thus improves the performance.

MICROSIGN EMPLOYER BRAND

The specially-abled friendly infrastructure and environment, which Microsign has created over the period through human engineering make Microsign employer of the choice for all abled and specially-abled people.

When asked about the attrition rate Nisheeth said that they have almost zero per cent attrition rate. Additionally, they have never given recruitment advertisements in the newspaper. All the employees have come through references.

Sama (2020) found that more than 90% of Microsign employees are highly satisfied in terms of differently-abled friendly infrastructure, salary and employee engagement activities. From giving the opportunity to specially-abled people and bringing them to mainstream business by discovering their abilities behind their disability, developing differently-abled friendly infrastructure, creating differently-abled friendly environment and culture without harming company's profit goals, Microsign has truly empowered differently-abled people. For this initiative, Microsign bagged many prestigious international and national awards like NCPDP Helen Keller Award (1999), National Award for Welfare of People with Disabilities (1999) from Government of India, ACMA (Automotive Component Manufacturers Association of India) (2019) to name a few. Thus, Microsign has worked as an eye-opener for those professional organizations who are ignoring or doubting the abilities of differently-abled people.

CORPORATE IMAGE, CORPORATE BRAND, CUSTOMER ACQUISITIONS AND RETENTIONS

Microsign philosophy is not only producing the quality products but also creating a better society by creating win–win situations for all stakeholders. The company proudly supply quality products with the efforts of 60% specifically-abled employees. Microsign believes in "live & let live" with dignity and honor. The company's visionary CSR policy has provided a positive shift in the social status of these employees. These social concern of the Microsign creates a good corporate image in the market.

Microsign ensures the sustainability of the environment by employing healthy practices such as using solar energy for water heating, rainwater harvesting to conserve groundwater, wind-solar hybrid generator as a renewable source of energy. The company has healthy HR policy concerning its stated key purpose of existence, viz. "to develop training and tools that help people bypass physical, mental, and personal limitations." This socially responsible image and environment concerned image strengthens the Microsign corporate image associations (Keller et al. 2011). It is in sync with the findings of Weiwei (2007) and Keller (2000).

Microsign's pious efforts received continues media attention. The success story of Microsign appeared in the leading national newspapers such as "No children of a lesser god," (Sunday Times, 2007); "Taking the less gifted along" (Ahmedabad Times, 2001); "Factory of fraternity" (DNA India, 2015); "Differently-abled to power Nano" (The Times of India, 2008). This media coverage projects the good corporate image of Microsign among all stakeholders.

These efforts of Microsign bagged it prestigious awards for excellence in quality and social work (Table 12.2). They adhere to the international quality standards, namely, ISO: 90027 and QS: 9000.8. Microsign's popularity is because of making international quality standard products with the contribution of 60% specially-abled employees. Nisheeth along with his one of the mentally challenged employees, Hardik Shah, got the opportunity to share their stories on a TV program hosted by celebrity such as Amitabh Bachchan. Due to this national and international awards and media coverage, Microsign emerged as a good corporate brand.

Covid-19 has impacted the Microsign business as an automobile sector registered zero sales in April 2020 due to stringent lockdown (Pandey, 2020). As automobiles manufacturers are their main clients, it has a drastic impact on the business. Due to this unprecedented crisis, the need for diversification arises to sustain in the business. Thus, Microsign attempted to offer their products to the pharmaceutical companies in Gujarat, India.

When asked how was the response from the pharma companies, Nisheeth took a deep breath and summarized the entire discussion by remembering how they got a red carpet in all big pharma companies, namely, Sun Pharma, Torrent Pharma, Cadila Pharma because of the Microsign corporate image and corporate brand in the market. He said, "Top officials of Sun Pharma immediately recognised the Microsign because of the social contribution and agreed to have empanelment with us for the supply of components." These customer acquisitions he attributed to good corporate image and brand of the Microsign.

Nisheeth said, "Their good corporate image and corporate brand not only help them in this crisis time in customer acquisition of pharma industries but also help them in retaining their existing clientele." He said with a smile that their existing clients (giant brands including Siemens, Larsen & Toubro Ltd., ABB; Bharat Dynamics Ltd., BHEL and ISRO, DRDO, Honda; Maruti, and Tata) give them always first preference to supply the components until their price is the same as their competitors. He said with a smile that they have zero percent customer defection. Thus, Microsign corporate brand is a competitive advantage and leads to customer retention.

CONCLUSION

Microsign underwent innovations in HR practices and initiated the hiring of specially-abled people way back in 1978 far before the government's disabilities act. Today Microsign has 60% of specially-abled employees serving quality products to the big multinational clients like Honda, Maruti, Tata, ISRO, Reliance Industries, to name a few. This social engineering made Microsign the employer of choice among abled and specially-abled people, resulting in good employer brand. Microsign's success stories featured in various popular media like Star Plus television show and in national newspapers like *Times of India*, *DNA*, *The Economic Times*, and *Indian Express*. They also received international and national awards like Hellen Keller Award. This creates a good corporate image of Microsign among all the stakeholders, resulting in good corporate brand. It helped the organization grow multifold not only in terms of financial returns but also in terms of its contribution to society (Pedersen, 2015). In the challenging times of Covid-19, Microsign corporate brand made by the legacy and heritage of hiring specially-abled and other CSR activities help them in customer acquisitions and retention.

CONTRIBUTION TO THEORY

This study contributes to the literature on employer branding through employment of hiring specially-abled people and CSR and its impact on corporate image and corporate brand in the perspective of SMEs. Through the case study of Microsign and semi-structured interviews of specially-abled people, this study unravels the impact of employer branding on corporate brand and its subsequent impact on customer acquisitions and retentions. The results found that employer branding through hiring specially-abled people creates positive corporate image and develops good corporate brand in the market, resulting in more customer acquisitions and retentions. There is limited literature to explore the connections among employer brand, corporate image and brand, and customer acquisitions and retention, especially from the perspectives of SMEs.

This research also explored the initiative of employers in hiring the specially-abled employees and their contribution to the organization success in the case of SMEs. This research established the positive impact of employer brand on corporate image and corporate brand. Further, it also established the benefits of good corporate image in acquisitions and retention of customers in challenging times like Covid-19.

CONTRIBUTION TO PRACTICE

From the perspective of managerial implications, this study established that employer branding by hiring specially-abled people has a positive impact on corporate image, leading to good corporate brand. This corporate brand helps a firm in acquiring and retaining the customers in challenging times like Covid-19. Taking a cue from this research, SMEs' employers can make a strategy to hire specially-abled employees to create win–win situation for specially-abled employees and organization.

Employers are exploring to develop a good employer and corporate brand through various activities. Taking a cue from this study, it is important to note that CSR and inclusive workforce are a key to employer branding. Hence, if employers want to make good corporate brand, then diverse workforce including specially-abled people can play a critical role in it. Further, it is also possible to acquired new customers by utilizing good corporate brand.

LIMITATIONS OF THE STUDY

This research has few limitations. This research focused on the impact of employer brand on customer acquisitions and retention in case of SME. Hence, the results cannot be generalized for MNCs and big companies. To add to the pool of existing literature, more studies should be conducted by considering other SMEs from metro cities. This case cited the evidence from a manufacturing firm; further studies should be conducted considering the service industries.

FUTURE DIRECTIONS

This research emphasizes on employer brand impact on customer acquisitions and retention in the case of an SME. Thus, future research can consider the MNCs and big firms. This study has not focused on the other aspects of employer branding like organizational culture and employee engagement activities. It will be important to explore the impact of these variables on employer branding and customer acquisitions and retention in the case of service firms.

REFERENCES

Aaker, D. A. (1996), "Measuring brand equity across products and markets", *California Management Review*, Vol. 38, No. 3, pp. 102–120.

Aaker, D. A. (2004), "Leveraging the corporate brand", *California Management Review*, Vol. 46, No. 3, pp. 6–18.

Ambler , T. and Barrow, S. (1996), "The employer brand", *Journal of Brand Management*, Vol. 4, No. 3 , pp. 185–206.

Andreassen, T. W., & Lindestad, B. (1998), "The impact of corporate image on quality, customer satisfaction and loyalty for customers with varying degrees of service expertise", *International Journal of Service Industry Management*, Vol. 9, No. 1, pp. 7–23.

Anselmsson, J. and Bondesson, N. (2013), "What successful branding looks like: A managerial perspective", *British Food Journal*, Vol. 115, No. 11, pp. 1612–1627.

Anselmsson, J., Bondesson, N., and Melin, F. (2016), "Customer-based brand equity and human resource management image", *European Journal of Marketing,* Vol. 50, No.7, pp. 1185–1208.

Backhaus, K. and S. Tikoo, (2004), "Conceptualizing and researching employer branding", *Career Development International*, Vol. 9, No. 5, pp. 501–517.

Barrow, S. and Mosley, R. (2005), *The Employer Brand*, London: Wiley.

Balmer, J. M. T. (2001), "Corporate identity, Corporate Branding and Corporate Marketing: Seeing through the Fog", *European Journal of Marketing*, Vol. 35, No. 3–4, pp. 248–291.

Bhadra, S., (2010), "Pre-requisites for Assessing Psychological Trauma among the Survivors of Disaster", *Psyber News: International Psychology Research Publication*. Vol. 1, No. 3, pp. 27–33. ISSN: 09760709.

Bhalla, J. (2015), "5 Lakh Differently-Abled To BE Skill-Ready by 18", *The Pioneer*, www.dailypioneer.com/2015/sunday-edition/5-lakh-differently-abled-to-be-skill-ready-by-18.html (accessed June 2019).

Bhatia, M. (2008), "IT firms give up a leg-up to the differently-abled", *The Economic Times*. https://economictimes.indiatimes.com/liveitup/it-firms-give-a-leg-up-to-the-differently-abled/articleshow/3806539.cms?from=mdr (accessed May 2020).

Bhattacharjee, S. (2011),"PWDs, an Untapped talent Pool", *The Times of India-Ascent*. www.timesascent.com/career-advice/PWDs-an-untapped-talent-pool/21195 (accessed December 2019).

Bruce, L. (2006) "Count me in: people with a disability keen to volunteer", *Australian Journal of Volunteering*, Vol. 11, No. 1, pp. 59–64.

Bruyere, S. M., Erickson, W. A., & VanLooy, S. A. (2006), "The impact of business size on employer ADA response". *Rehabilitation Counseling Bulletin*, Vol. 49, No. 4, pp. 194–206.

Bucharest, H. B., Keys, C., and Balcazar, F. (2000), "Employer attitudes toward workers with disabilities and their ADA employment rights: A literature review", *Journal of Rehabilitation*, Vol. 66, No. 4, pp. 4–16.

Cable, D. M. and Graham, M. E. (2000), "The determinants of job seekers' reputation perceptions", *Journal of Organizational Behavior*, Vol. 21, No. 8, pp. 929–947.

Chernatony, L. (1999), "Brand management through narrowing the gap between brand identity and brand reputation", *Journal of Marketing Management*, Vol. 15, No. 1&3, pp. 157–179.

Chernatony, L. (2006), *From Brand Vision to Brand Evaluation. The strategic process of growing and strengthening brands*, 2nd edn., Oxford: Elsevier LTD.

Collins, J. (2001), *Good to Great*, London: Random House.

Commonwealth Department of Social Services (2011), "National Disability Strategy 2010–2020", Canberra, Australia [online]. Available at: www.dss.gov.au/ourresponsibilities/disability-andcarers/publications-articles/policy-research/nationaldisability-strategy-2010-2020 (accessed August 2020).

Csillag, S. and Gyori, Z. (2016), "Is there a place for me?" Employment of people with disabilities as part of CSR strategy", *Proceedings of the 4h Strategica International Academic Conference: Opportunities and risks in the contemporary business environment*, ISBN: 978-606-749-181-4, 20-21 October 2016, Bucharest, Romania, 860–872.

Csillag, S. and Győri, Z. (2016), "Is there a place for me? Employment of PWD as part of CSR strategy", In *Strategic International Academic Conference* (Vol. 4, pp. 860–872).

Dawkins, P. and Reichheld, F. (1990), "Customer retention as a competitive weapon", *Directors and Boards*, Vol. 14, No. 4, pp. 42–47.

DEOC (Diversity and Equal Opportunity Centre), (2009), "Employment of Disabled People in India", Baseline Report, Prepared for National Centre for Promotion of Employment for Disabled People (NCPEDP).

Dibben, P., James, P. & Cunningham, I. (2002) Senior management commitment to disability: the influence of legal compulsion and best practice, *Personnel Review*, Vol. 30, pp. 454–467.

Du Preez, R., and Bendixen, M. T. (2015), "The impact of internal brand management on employee job satisfaction, brand commitment and intention to stay", *International Journal of Bank Marketing*, Vol. 33, No. 1, pp. 78–91.

Ewing, M. T., Pitt, L. F., De Bussy, N. M., and Berthon, P. (2002), "Employment branding in the knowledge economy", *International Journal of Advertising*, Vol. 21, No. 1, pp. 3–22.

Fasciglione, M. (2015), "Article 27 of the CRPD and the right of inclusive employment of people with autism", In *Protecting the Rights of People with Autism in the Fields of Education and Employment* (pp. 145–170). Cham: Springer.

Fombrun, C. and Shanley, M. (1990), "What's in a name? Reputation building and corporate strategy", *Academy of Management Journal*, Vol. 33, pp. 233–56.

Fombrun, C. J. and Gardberg, N. (2000), "Who's tops in corporate reputation?", *Corporate Reputation Review*, Vol. 3, No. 1, pp. 13–17.

Gaddam, S. (2008), "Modeling employer branding communication: The softer aspect of HR marketing management", *The ICFAI Journal of Soft Skills*, Vol. II, No. 1, pp. 45–56.

Ghosh, L.(2011), "Companies Unwittingly Discriminate Against Differently Abled", *The Economic Times*, Available at: https://economictimes.indiatimes.com/jobs/companies-unwittingly-discriminate-against-differently-abled/articleshow/7870126.cms?from=mdr (accessed May 2020).

Gioia, D. A., Schultz, M., & Corley, K. G. (2000), "Organizational identity, image, and adaptive instability", *Academy of Management Review*, Vol. 25, No.1, pp. c63–81.

Goyal, M. (2013), "SMEs employ close to 40% of India's workforce, but contribute only 17% to GDP", *The Economic Times*, Available at: https://economictimes.indiatimes.com/small-biz/policy-trends/smes-employ-close-to-40-of-indias-workforce-but-contribute-only-17-to-gdp/articleshow/20496337.cms (accessed Jun 2020)

Graffam, J., Smith, K., Shinkfield, A. and Polzin, U. (2002) "Employer benefits and costs of employing a person with a disability", *Journal of Vocational Rehabilitation*, Vol. 17, No. 4, pp. 251–263.

Greening, D. W. and Turban, D. B. (2000), "Corporate social performance as a competitive advantage in attracting a quality workforce", *Business and Society*, Vol. 39, No. 3, pp. 254–280.

Griffin, P., Peters, M. L., and Smith, R. M. (2007) *Ableism curriculum design In Teaching for diversity and social justice*, pp. 359–382. Routledge.

Griggs, L. B. (1995), "Valuing diversity: Where from... Where to?", In Griggs, L. B. & Louw, L. L. (Eds.), *Valuing Diversity: New Tools for a New Reality*. New York: McGraw-Hill.

Gröschl, S. (2005), "Persons with disabilities: A source of nontraditional labor for Canada's hotel industry", *Cornell Hotel and Restaurant Administration Quarterly*, Vol. 46, No. 2, pp. 258–274.

Hatch, M. J. and Schultz, M. (1997), "Relations between organizational culture, identity and image", *European Journal of Marketing*, Vol. 31, No. 5/6, pp. 356–365.

Hatch, M. J. and Schultz, M. (2001), "Are the strategic stars aligned for your corporate brand?" *Harvard Business Review*, February, Vol. 79, No. 2, pp. 128–134.

Hatch, M. J., & Schultz, M. (2003)," Bringing the corporation into corporate branding", *European Journal of Marketing*, Vol. 37, No. 7/8, pp. 1041–1064.

Helm, S. (2013), "A matter of reputation and pride: Associations between perceived external reputation, pride in membership, job satisfaction and turnover intentions", *British Journal of Management*, Vol. 24, No. 4, pp. 542–556.

Herrbach, O. and Mignonac, K. (2004), "How organisational image affects employee attitudes", *Human Resource Management Journal*, Vol. 14, No. 4, pp. 76–88.

Hernandez, B., McDonald, K., Divilbiss, M., Horin, E., Velcoff, J., and Donoso, O. (2008), "Reflections from employers on the disabled workforce: Focus groups with healthcare, hospitality and retail administrators", *Employee Responsibilities and Rights Journal*, Vol. 20, No. 3, 157–164. https://doi.org/10.1007/s10672-008-9063-5

Hidegh, A. L., and Csillag, S. (2013), "Toward 'mental accessibility': Changing the mental obstacles that future Human Resource Management practitioners have about the employment of PWD", *Human Resource Development International*, Vol. 16, No. 1, 22–39. https://doi.org/10.1080/13678868.2012.741793

Jamka, B. (2011), "Wartości a model biznesowyzarządzaniaróżnorodnością", *Master of Business Administration*, Vol. 19, 6, pp. 65–75.

Keller, K. L. (1993), "Conceptualizing, measuring, and managing customer-based brand equity", *Journal of Marketing*, Vol. 57, No. 1, pp. 1–23.

Keller, K. L. (2000) *Building and managing corporate brand equity. The Expressive Organization*. Oxford: Oxford University Press, pp. 116–137.

Keller, K. L., Parameswaran, M. G., and Jacob, I. (2011). *Strategic Brand Management: Building, Measuring, and Managing Brand Equity*. India: Pearson Education.

Kennedy, S. H. (1977), "Nurturing corporate images", *European Journal of Marketing*, Vol. 11, No. 3, pp. 119–164.

Kapferer, J. N. (2004), *The New Strategic Brand Management*. London: Kogan Page.

Kimpakorn, N. and Tocquer, G. (2009), "Employees' commitment to brands in the service sector: Luxury hotel chains in Thailand", *Journal of Brand Management,* Vol. 16, pp. 532. https://doi.org/10.1057/palgrave.bm.2550140

Knowles, J., Ettenson, R., Lynch, P., and Dollens, J. (2020), "Growth opportunities for brands during the COVID-19 crisis", *MIT Sloan Management Review*, Vol. 61, No. 4, 2–6.

Kuznetsova, P., Ordonez, V., Berg, A., Berg, T., and Choi, Y. (2012, July), "Collective generation of natural image descriptions", In *Proceedings of the 50th Annual Meeting of the Association for Computational Linguistics (Volume 1: Long Papers)* (pp. 359–368).

Lewis, G. and Priday, J. (2008) National Disability Partnership Project: Preliminary Findings, unpublished paper, EDGE Employment Solutions Inc., Perth, WA, Centre for Research into Disability and Society, Curtin University of Technology, Bentley, WA, Group Training Australia Ltd, Sydney, with funding from the Australian Government Department of Education, Employment and Workplace Relations, Canberra.

Lindsay, S., Cagliostro, E., Albarico, M., Mortaji, N., and Karon, L. (2018), "A systematic review of the benefits of hiring people with disabilities", *Journal of occupational rehabilitation*, Vol. 28, No. 4, pp. 634–655.

Lloyd, S. (2002), "Branding from the inside out", *Business Review Weekly*, Vol. 24, No. 10, pp. 64–66.

Martin, G., Beaumont, P., Doig, R., and Pate, J. (2005), "Branding: A new performance discourse for HR?", *European Management Journal*, Vol. 23, No. 1, pp. 76–88.

Matuska, E. and Sałek-Imińska, A. G. N. I. E. S. Z. K. A. (2014). Diversity management as employer branding strategy—Theory and practice. *Human Resources Management & Ergonomics*, Vol. 8, No. 2, pp. 72–87.

McQueen, A. (2004), "Suds Law", In *The King of Sunlight*. London: Bantam Press.

Microsign. (2018). www.microsignproducts.com. Retrieved from www.microsignprodu- cts. com/clients/ (accessed November 2019).

Miethlich, B. and Šlahor, Ľ. (2018), "Employment of persons with disabilities as a corporate social responsibility initiative: Necessity and variants of implementation", In *Innovations in Science and Education, CBU International Conference, Prague, 21–23 March 2019* (pp. 350–355). Prague: CBU Research Institute.

Miethlich, B., and Oldenburg, A. (2019), "How social inclusion promotes sales: An analysis of the example of employing people with disabilities", *Journal of Marketing Research and Case Studies*, 1–15.

Minchington, B. (2005), *Employer Brand Leadership–. A Global Perspective.* Australia.

Mitra, S., and Sambamoorthi, U. (2006), "Employment of persons with disabilities: Evidence from the National Sample Survey", *Economic and Political Weekly*, pp. 199–203.

Mokina, S. (2014), "Place and role of employer brand in the structure of corporate brand", *Economics & Sociology*, Vol. 7, No. 2, 136.

Monachino, M. S. and Moreira, P. (2014), "Corporate social responsibility and the health promotion debate: An international review on the potential role of corporations", *International Journal of Healthcare Management*, Vol. 7, No. 1, 53–59. https://doi.org/10.1179/2047971913y.0000000058

Moore, D. and Fishlock, S. (2006), Can do Volunteering: A Guide to Involving Young Disabled Volunteers [online]. Available at: www.energizeinc.com/art/subj/documents/canDOweb.pdf (accessed November 2019).

Moroko, L. and Uncles, M. D. (2008), "Characteristics of successful employer brands", *Journal of Brand Management*, Vol. 16, No. 3, pp. 160–175.

Mosley, R. W. (2007), "Customer experience, organisational culture and the employer brand", *Journal of Brand Management*, Vol. 15, No. 2, 123–134.

Mosley, R. (2009), "Employer brand. The performance driver no business can ignore". www.marksherrington.com/downloads/Richard%20Mosley%20eArticle.pdf (accessed 10 March 2013).

NCPEDP (1999), "Employment practices of the corporate sector. Delhi: NCPEDP". Available from www.ncpedp.org/employ/emresrch.htm (accessed May 2020).

Pandey, A. (2020), "Covid-19 impact: Automakers' domestic sales draw a blank in 2020". www.livemint.com/companies/people/covid-19-impact-automakers-domestic-sales-draw-a-blank-in-april-11588335477154.html

Pappu, R., Quester, P. G., and Cooksey, R. W. (2007), "Country image and consumer-based brand equity: Relationships and implications for international marketing", *Journal of International Business Studies*, Vol. 38, No. 5, pp. 726–745.

Pedersen, E. R. G. (Ed.). (2015). *Corporate social responsibility*. Los Angeles, CA: Sage Publications.

Priyadarshi, P. (2011), "Employer brand image as predictor of employee satisfaction, affective commitment & turnover", *Indian Journal of Industrial Relations*, Vol. 46, No. 3, pp. 510–522.

Rana, G. and Rastogi, R. (2015), "Organizational justice enhancing managerial effectiveness in terms of activity of his position, achieving results and developing further potential", *Research on Humanities and Social Sciences*, Vol. 5, No. 1, 24–31.

Rana, G., Rastogi, R. and Garg, P. (2016), "Work values and its impact on managerial effectiveness: A relationship in Indian context", *Vision*, Vol. 22, pp. 300–311.

Rana, G. and Sharma, R. (2019), "Assessing impact of employer branding on job engagement: A study of banking sector", *Emerging Economy Studies*, Vol. 5, No. 1, 7–21. https://doi.org/10.1177/2394901519825543

Rana, S. and Sharma, R. (2018), "An overview of employer branding with special reference to Indian organizations", In Sharma, N., Singh, V. K., and Pathak, S. (Eds.), *Management Techniques for a Diverse and Cross-Cultural Workforce* (pp. 116–131). Hershey: IGI Global. http://doi:10.4018/978-1-5225-4933-8.ch007.

Rana, R. E. N. U., and Kapoor, S. H. I. K. H. A. (2016), "Exploring the contribution of employer branding in corporate image building", *International Journal of Business and General Management*, Vol. 21–32, 37–42 [Special Edition].

Rangan, P. (2011), "More PWDs in Today's Workforce", *The Hindu*. Available at: www.thehindu.com/news/cities/Hyderabad/more-pwds-in-todays-workforce/article2539958.ece (accessed May 2020).

Reichheld, F. F. (1996). Learning from customer defections. *Harvard Business Review*, Vol. 74, No. 2, pp. 56–67.

Reichheld, F. F. (1996) *The Loyalty Effect*. Boston: Harvard Business School Press.

Ruhindwa, A. (2016) "Exploring the challenges experienced by people with disabilities in the employment sector in Australia: advocating for inclusive practice-a review of literature", *Journal of Social Inclusion*, Vol. 7, No. 1, pp. 4–19.

Sama, R. (2020), "Envisioning a mutually inclusive growth story: A case study of Microsign products. www.inderscience.com/info/ingeneral/forthcoming.php?jcode=ijttc.

Sharma R., Singh S. P., and Rana G. (2019). "Employer branding analytics and retention strategies for sustainable growth of organizations", In Chahal, H., Jyoti, J., and Wirtz, J. (Eds.), *Understanding the Role of Business Analytics*. Singapore: Springer. https://doi.org/10.1007/978-981-13-1334-9_10.

Siperstein, G. N., Romano, N., Mohler, A., and Parker, R. (2006), "A national survey of consumer attitudes towards companies that hire people with disabilities", *Journal of Vocational Rehabilitation*, Vol. 24, No. 1, pp. 3–9.

Somvanshi, K. K. (2015), "India inc still a challenge for disabled, 10 cos employ 90% of disabled employees working in nifty 50 firms", *The Economic Times*. Available from: https://economictimes.indiatimes.com/jobs/india-inc-still-a-challenge-for-disabled-10-cos-employ-90-ofdisabled-employees-working-in-nifty50-firms/articleshow/50019057.cms (accessed June 2020).

Statista (2011). Available at: www.statista.com/statistics/615781/households-by-size-india/ (accessed June 2020).

Suter, R., Scott-Parker, S., and Zadek, S. (2007), *Realising potential: Disability confidence builds better business*. Ithaca, NY: Cornell University Press.

Tamberg, V., Badyin, A. (2009), PR strategy. The formula of corporate image, Available from: http://newbranding.ru/articles/4r-public-relations-strategy (accessed June 2020).

Thomas, A. and Lahla, M. A. (2002) "Factors that influence the employment of people with disabilities in South Africa", *South African Journal of Labour Relations*, Vol. 26, No. 4, pp. 4–32.

Trivedi, J. (2020), "Effect of corporate image of the sponsor on brand love and purchase intentions: The moderating role of sports involvement", *Journal of Global Scholars of Marketing Science*, Vol. 30, No. 2, pp. 188–209.

Turnbull, P. W., and Wilson, D. T. (1989), "Developing and protecting profitable customer relationships", *Industrial Marketing Management*, Vol. 18, No. 3, 233–238.

Urde, M. (2013), "The corporate brand identity matrix", *Journal of Brand Management*, Vol. 20, No. 9, pp. 742–761.

Waterhouse, P., Kimberley, H., Jonas, P., and Glover, J. (2010), What Would it Take? Employer Perspectives on Employing People with a Disability, National Centre for Vocational Education Research, Adelaide, South Australia.

Weiwei, T. (2007), "Impact of corporate image and corporate reputation on customer loyalty: A review", *Management Science and Engineering,* Vol. 1, No. 2, 57–62.

Wilden, R., Gudergan, S., Lings, I. (2010), "Employer branding: Strategic implications for staff recruitment", *Journal of Marketing Management*, Vol. 26, No. 1–2, pp. 56–73.

Yeend, P. (2012), Budget 2011–2012: Disability Support Pension-Reforms, Parliament of Australia, Canberra, Australia [online]. Available at: www.aph.gov.au/about_parliament/ Parliamentary_ departments/parliamentary_library/pubs/rp/budgetreview201112/ disability (accessed July 2020).

Index